R Companion to Elementary Applied Statistics

T0186906

R Companion to Elementary Applied Statistics

Christopher Hay-Jahans

CRC Press
Taylor & Francis Group
Boca Raton London New York

CRC Press is an imprint of the
Taylor & Francis Group, an **informa** business

A CHAPMAN & HALL BOOK

CRC Press
Taylor & Francis Group
6000 Broken Sound Parkway NW, Suite 300
Boca Raton, FL 33487-2742

© 2019 by Taylor & Francis Group, LLC
CRC Press is an imprint of Taylor & Francis Group, an Informa business

No claim to original U.S. Government works

Printed on acid-free paper

International Standard Book Number-13: 978-1-138-32916-4 (Paperback)
978-1-138-32925-6 (Hardback)

Library of Congress Cataloging-in-Publication Data

Names: Hay-Jahans, Christopher, author.
Title: R companion to elementary applied statistics / Christopher Hay-Jahans.
Other titles: Companion to elementary applied statistics.
Description: Boca Raton : CRC Press, Taylor & Francis Group, 2018. | Includes bibliographical references and index.
Identifiers: LCCN 2018030262 | ISBN 9781138329164 (pbk.).
Subjects: LCSH: Statistics–Data processing. | R (Computer program language).
Classification: LCC QA276.45.R3 H39 2018 | DDC 519.50285/5133--dc23
LC record available at https://lccn.loc.gov/2018030262

**Visit the Taylor & Francis Web site at
http://www.taylorandfrancis.com**

**and the CRC Press Web site at
http://www.crcpress.com**

*To my family
for their support and patience,
and to my students
who inspire me to learn new ideas
through their curiosity,
enthusiasm, and willingness to learn.*

Contents

Preface

This work includes traditional applications covered in elementary statistics courses as well as some additional methods that address questions that might arise during or after the application of commonly used methods. No familiarity with R or programming background is required to begin using this book, and while reasonable quantitative skills will prove useful, a minimal formal background in statistical theory should suffice.

Theoretical foundations are not treated; however, relevant assumptions and quantitative background information, common rules of thumb, interpretations of results from methods covered, and recommendations from the literature are discussed as needs arise. Since the emphasis is on computational steps and coding associated with methods presented, the methods discussed are used to introduce the fundamentals of using R functions and provide ideas for developing further skills in writing R code. These ideas are illustrated through an extensive collection of examples.

Beginning with basic tasks and computations with R, readers are then guided through ways to bring data into R, manipulate the data as needed, perform common statistical computations and elementary exploratory data analysis tasks, prepare customized graphics, and take advantage of R for a wide range of methods that find use in many elementary applications of statistics. The R tools used are almost all that come with a default/base installation of the program. These include the packages `stats`, `graphics`, `grDevices`, `utils`, `datasets`, `methods`, and `base` [115]. While some examples of installing and drawing from other specialized R packages appear as needs arise, it will be found that the base packages are almost always adequate for elementary applications.

About Using This Companion

This book can be used for two purposes, particularly for individuals interested in learning how to use R. It can be used as a resource for a project-based elementary applied statistics course through a selection of sections. Or, it can serve as a resource for professionals and researchers whose work involves using methods covered in this book, and who might wish to delve (a little more deeply) into using R.

Quite a few methods presented may be considered non-traditional, or advanced, as they do not appear in texts at the elementary statistics level. For the interested reader, references containing theoretical and other relevant details are cited for these methods as they appear. In cases where sources/references are not cited (mostly for commonly encountered traditional elementary statistical tools/methods) the reader is referred to any one of the many elementary statistics texts available, for example, [12], [40], [77], or [95] or, if so desired, more theoretically inclined works such as [7] or [69].

There are some supplements available for users of this book that are posted on the accompanying CRC website (use the ISBN: 9781138329164), which can be found at

http://www.crcpress.com

This website contains a README file with further details that the reader will find helpful in getting the most out of this book. Also posted are the accompanying script files for all examples presented in the book, as well as extensions of code presented including illustrations of preparing fairly complex code. The data files needed for examples presented are contained on this website.

About the Code

The code contained in this book (and the accompanying script files) were prepared and run using R Version 3.4.4 [115] in the Windows environment. However, there should be no worries about using the same code with newer versions of R, and on a Linux or Macintosh OS X platform. While navigating around the three systems may differ in many ways, the actual R code is, for the most part, consistent across all three. In the case of using R on a Macintosh, the R-Project website is a good first resource for frequently asked questions (and answers). In all cases, internet searches reveal many resources to bridge the rare differences between operating R in the Windows, OS X and Linux environments.

The basic R Editor and RGui have intentionally been used for this book. For those who might want a platform that contains more features, an alternative open source platform that is gaining popularity is *RStudio*. Code presented in this book will work in RStudio, and documentation and tutorials on using RStudio are available on the RStudio website.

About the Data Used

Almost all of the data used for illustrations were simulated for specific purposes. The methods illustrated can be extended to real data (or a different story) as needed. For those interested in gaining access to open source data, one source is package datasets. This is one of the default packages loaded on installing R, and it contains a fairly large and varied collection. A complete list of datasets in this package, along with brief descriptions and links to documentation pages, can be obtained by running ?"datasets" in the console. The function call data() provides information on accessing further such packages. Other packages in CRAN that contain datasets will need to be installed and then loaded into the workspace before any included datasets can be accessed, see Section 1.8.

The R community has gathered quite a bit of information on finding data on the internet and a web-search (typically with keywords such as "R" and "Free Data") will reveal sites that list links to open source data sets. For example, at the time of writing these lines, the following page

https://r-dir.com/reference/datasets.html

was found on the *R Software Reference Directory*. This site contains links to a wide variety of public domain data sources. Any number of datasets from these sources may be imported using methods described in Chapter 2, and explored using methods covered in later chapters.

Acknowledgments

Many thanks are due to the CRC reviewers for their valuable observations and suggestions for improving the exposition in this work, and to the CRC publishing team for providing a friendly, helpful and extremely efficient (and painless) process from proposing to publishing. I would also like to thank Dr. Karl Moder of the University of Natural Resources and Life Sciences, Vienna, for sharing his R coding ideas in connection to Hotelling's T^2 test. Ideas for the content of this book, particularly on extensions in coding, arose from a wide range of questions raised by my students, so they are due thanks for keeping me on my toes.

This book was prepared using Scientific Workplace, Version 5.5, and thanks are due to George Pearson of MacKichan Software, Inc. for his very prompt help in answering typesetting questions.

1

Preliminaries

Combined with the Preface, this chapter contains information on how ideas and material are presented in this book. So, all readers are encouraged to, at least, browse through the contents. Included are the basics needed to get started, some terminology, brief introductions to data structures, operators and functions, on preparing R code, and on getting help on the use of R.

1.1 First Steps

For starters, R can be downloaded from the *R Project for Statistical Computing* website, located at

```
http://www.r-project.org/
```

Once on this website, follow the directions given in the "Getting Started" page. The process of installing R is surprisingly simple and quick. Once installed, R can be started up in the usual manner by clicking on the R icon, the initial window that appears when R is started up is shown in Figure 1.1.

Before moving on, take some time to familiarize yourself with the contents of the menu items located in the *RGui toolbar*. At start-up, the *R Console* also provides some quick introductory information on R.

To quit R, either run q() in the console or use the **File→Exit** sequence in the RGui toolbar. When quitting R after a session covering any part of this book it is suggested that the "No" button be clicked on in response to the question "Save workspace image?" in the pop-up window. This ensures unnecessary clutter does not get reloaded in a later session.

1.2 Running Code in R

In this book, two environments are used to run code in: the R Console can be used to run lines of code directly; and the R Editor can be used to first prepare, proofread and then run code.

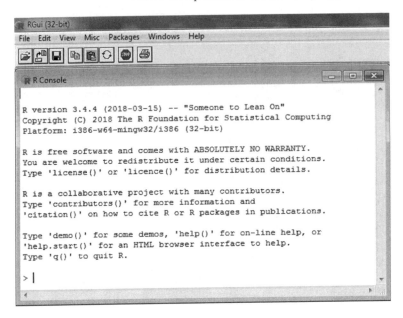

FIGURE 1.1
The opening window of the R Console showing the RGui toolbar and some preliminary information.

The R Console

Code can be typed directly in the *R Console* through code entered at the *cursor* (after the > symbol) and then run by pressing the Enter key on the keyboard. For example, run each line of code shown below.

```
> data(cars)
> names(cars)
[1] "speed" "dist"
> plot(cars)
> summary(cars)
      speed           dist
Min.   : 4.0   Min.   :  2.00
1st Qu.:12.0   1st Qu.: 26.00
Median :15.0   Median : 36.00
Mean   :15.4   Mean   : 42.98
3rd Qu.:19.0   3rd Qu.: 56.00
Max.   :25.0   Max.   :120.00
```

Here are brief descriptions of what tasks have been performed.

The code in the first line is a *function call*, of the *function* data, which loads a dataset called cars into the *workspace*. The second line prints the names of the contents of cars which appear on the third line. The fourth line constructs a scatterplot of the data contained in cars (which shows up in an *R Graphics window*, figure not shown), and the fifth line computes some common summary statistics of the data in cars. The remaining lines are output produced by running the code summary(cars).

Typical output from running code presented in this book appears either in a graphics device window, such as for the code in line 4, or as lines in the console that are not preceded by a ">" or a "+", or as lines preceded by a bracketed number, for example, lines 3, and 6 through 12.

There are occasions when a function call for a particular task may be too long to fit neatly on a single line. In such cases the enter key can be pressed at any point in the code to move the cursor to a new line, at this point a "+" appears on the new line. A "+" is automatically placed at the start of a new line whenever the code in the preceding line is considered incomplete by the *R interpreter*. For example, type each of the following lines in the console one by one, pressing the Enter key after each line.

```
> plot(cars, xlab = "Speed (mph)",
+ ylab = "Stopping distance (ft)",
+ main = "Stopping Distance vs. Vehicle Speed")
```

The R interpreter is *case sensitive*. That is, code should be entered exactly as it is defined in order to be interpreted and run. For example, in running

```
> Plot(cars)
Error: could not find function "Plot"
```

an incorrectly named function is called, and in

```
> qnorm(P = c(0.05, 0.95))
Error in qnorm(P = c(0.05, 0.95)) :
unused argument (P = c(0.05, 0.95))
```

using upper-case "P" in place of lower-case "p" in the code results in a failure in the interpretation of the code entered.

Entering code in the console is convenient for certain tasks, mainly those that involve quick computations or checks that do not involve lengthy code. There is an alternative environment within which lengthy code can be prepared and checked before running it.

```
R Z:\Docs\AppStats\ScriptFiles\Chapter1\firstRun.R - R Editor
# Load the dataset cars
data(cars)
# Get the names of the contents of cars
names(cars)
# Plot the data in cars
plot(cars)
# Obtain the summary of the contents of cars
#
# An example of multi-line code
plot(cars, xlab = "Speed (mph)",
     ylab = "Stopping Distance (ft)",
     main = "Stopping Distance vs. Vehicle Speed")|
```

FIGURE 1.2
Example of a script file opened in the R Editor. Lines that begin with a "#"
symbol are comment lines, included to add clarity to code. Function calls that
extend to more than one line are indented to add clarity to code. See Section
1.10 for the indentation conventions used in this book.

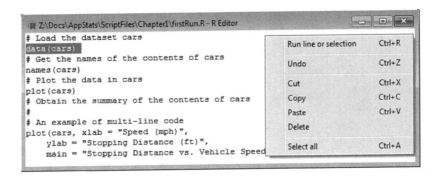

FIGURE 1.3
To run a single line of code in the R Editor, highlight or right-click on the
line of interest and select **Run line or selection**, or use the key sequence
Ctrl+R.

Script Files and the R Editor

The same tasks performed previously in the console can be accomplished
in the *R Editor* by first preparing a *script file* in the editor and then running
the code. Here is how this is done.

First open a new script file using the sequence **File→New script** in the
RGui toolbar (or **Crtl+N** on the keyboard). Next, type in the code as shown
in Figure 1.2; note that a "+" is not placed at the start of new lines in multi-line
function calls when preparing code in the R Editor.

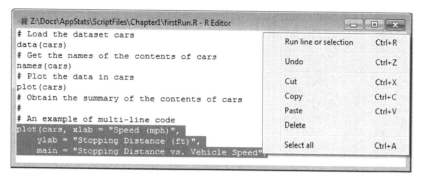

FIGURE 1.4

To run more than one line of code in the R Editor, highlight the lines of interest, then right-click the mouse and select **Run line or selection**, or use the keystroke sequence **Crtl+R**.

Code in the R Editor can then be run line-by-line as shown in Figure 1.3, or multiple lines of code can be run as shown in Figure 1.4.

Running the code as described in Figures 1.3 and 1.4 duplicates the tasks performed previously in the console,[1] and the same code (plus comment lines, if run) appear on the console. Such files can then be saved for later recall using the toolbar sequence **File→Save as...**, then selecting a suitable folder and giving the file a name. Script files in R are given the extension ".R", for example, this file could be called `testRun.R`.

All code used in this book, including additional exploratory exercises, are contained in script files for each chapter. For example, this chapter's code are stored in the file `Chapter1Script.R`. Close the script file just created, then open the file `Chapter1Script.R` using the sequence **File→Open script** in the RGui toolbar (or **Crtl+O** on the keyboard).[2]

1.3 Some Terminology

This section introduces R-related terminology that appears in this book. For a more complete discussion on technical details and terminology associated with R and the objects it works on, readers are referred to the user manuals [114] and [136] that are downloaded during the installation process (use the RGui menu sequence , **Help→HTML help...**).

---

[1]For Macintosh users, the keystroke equivalent of **Crtl+R** is **Command+Enter**.

[2]It is useful to tile the R Console and R Editor windows when using a script file, for example, use the RGui toolbar sequence **Windows→Tile Vertically**.

The term *object* refers to any specialized structure on which the R interpreter operates. Each object belongs to a particular *class*, and is of a specific *type*. Here are some examples.

First create an object by running

```
> (x <- c(0, 1))
[1] 0 1
```

Here is an explanation of what just happened. The *combine function*, c, gathers the numbers 0 and 1 in a single object to which the *assignment operator*, <-, assigns the name x.[3]

By enclosing an *assignment statement* within parentheses, the R interpreter is instructed to print the result of this operation to the screen. If the assignment were not to be enclosed by parentheses, then the contents of object x would not be printed to the screen. To display the contents of x on the screen (again), simply enter x.

Code for two or more separate evaluations can be placed on the same line if they are separated by a semicolon. So, for example, to test whether x is a vector, and determine its class and type, run

```
> is.vector(x); class(x); typeof(x)
```

The objects is.vector, class, and typeof are further examples of object classification/testing functions. As will be seen in running these function calls, the output indicates that x is a double-precision numeric vector. There are other object testing functions like is.vector available to test whether a given object is of a certain class or type, here is a sampling.

is.vector	is.factor	is.list
is.numeric	is.integer	is.character
is.na	is.nan	is.null

The names of these functions are fairly self-explanatory, except possibly for is.na (determines if an object, or entries of an object are not available), is.nan (determines if an object, or entries of an object are not numbers) and is.null (determines if an object, or entries of an object are empty).

Here are some more examples of objects that will be encountered. To find out about the earlier loaded object cars, run

```
> class(cars); typeof(cars)
```

---

[3]Equivalently, -> or = can be used. For example, c(0, 1) -> x and x = c(0, 1) perform essentially the same task.

The output will show that `cars` is of class *data frame*, and type *list*. Furthermore, as seen earlier, running `names(cars)` shows that `cars`, itself, contains two objects named `speed` and `dist`.

Finally, consider an object such as `summary` on its own, and when it is used to perform an evaluation. First, take a look at the class and type of this object,

```
> class(summary); typeof(summary)
```

and then look at the class and type of the object that this function creates.

```
> carsSummary <- summary(cars)
> class(carsSummary); typeof(carsSummary)
```

The output from these runs inform us that `summary` is a function of type closure (a type specific to functions). But, `carsSummary`, the object created by applying `summary` to `cars`, is a table of type character.

1.4 Hierarchy of Data Classes

As a general rule, vectors of different classes should not be combined into a single vector. However, this does not necessarily mean that they cannot be combined in R. Here are illustrations of what happens when vectors of the three main classes used in this book are combined.

First create three vectors, one of class *logical*, the next of class *numeric*, and the third of class *character*.

```
> a <- c(TRUE, FALSE); class(a)
[1] "logical"
> b <- c(2.1, 3.5); class(b)
[1] "numeric"
> c <- c("A", "B"); class(c)
[1] "character"
```

Now use the combine function, c, to see what happens with various combinations of these three vectors by running

```
> c(a, b); class(c(a, b))
> c(a, c); class(c(a, c))
> c(b, c); class(c(b, c))
```

As will be seen from the output, if a character vector is combined with either of the others, the result is a character vector. If a numeric vector is combined with a logical vector, the result is a numeric vector.

1.5   Data Structures

In this book, the term *data structure* (or simply data) refers to any R object that contains data (in the traditional sense) to be manipulated or analyzed. The contents of such structures may be qualitative, quantitative, logical, or various combinations. The four most commonly used data structures in this book are *vectors*, *factors*, *data frames*, and *lists*. Two others that occur periodically are *matrices* and *arrays*. This section provides brief introductions to these structures; further details appear in later chapters where applications arise.

Vectors

The term *vector* will refer to any object containing a set of entries, say $\{x_1, x_2, \ldots, x_n\}$, where the x_i, $i = 1, 2, \ldots, n$ are all of the same class (numeric, character, or logical).

A univariate sample of quantitative data is typically stored as a *numeric vector*. For example,

```
> weights <- c(2.5, 3, 1.4, 2.7)
```

and a summary of `weights` can be obtained using the `summary` function. Similarly, logical data are typically stored in a *logical vector*. For example,

```
> good <- c(TRUE, FALSE, TRUE, TRUE)
```

and a summary of this vector produces a frequency distribution of the logical values contained in `good`. As is evident from the summary of `good`, there are two logical values that might appear in a vector; `TRUE` or `FALSE`. The value `NA` represents a value that is Not Available, and is typically used to flag missing data. While `NA` is itself a logical constant, it can be placed in vectors of any of the three types mentioned here, without any coercion of types occurring.

Finally, qualitative data can be stored as a *character vector*. For example,

```
> colors <- c("Red", "Red", "Blue", "Red")
```

As with the previous two vectors, a summary can be obtained using the `summary` function. However, it will be noticed that the resulting summary is not useful in the statistical sense. This leads to the next data structure.

Factors

The vector `colors` can be redefined as a *factor* using the code

```
> colors <- as.factor(colors)
```

and then summarized using the **summary** function. In this case the summary appears as a frequency distribution of the contents of **colors**. Any vector can be redefined as a factor and, in this book, categorical data are always defined as factors if any form of statistical analyses are to be performed on the data.[4]

In this book the term *variable* is used interchangeably with the terms vector or factor when dealing with data.

Data Frames

The previously loaded data structure **cars** is an example of a data frame. When printed to the console, data frames appear in table format where the columns contain vectors or factors. As with vectors and factors, a summary of the contents of a data frame can be obtained using the **summary** function. While the column lengths of vectors (and factors) in a given data frame need to be equal, the classes of the vectors placed in a data frame need not be the same.

Lists

Consider, for example, a study in which random samples of the ages of female lawyers from four Midwestern cities are obtained. One way in which these data can be placed in R is to store the samples in a *list*, say **lawyerAges** (see script file). A summary of the contents of a list of vectors (or factors) can be obtained using, for example,

```
> sapply(X = lawyerAges, FUN = summary)
```

The vectors (and/or factors) in a list do not need to have the same lengths and, when printed to the console, lists do not appear in table format. This example demonstrates just one simple use of lists. In fact, a list can contain objects of a wide range of classes and types.

Matrices and Arrays

A *matrix* is a 2-dimensional data structure with its contents being identified by rows and columns. Consider, for example, the following 3×4 matrix.

```
> (x <- matrix(data = c(1,  2,  3,  4,
+                       5,  6,  7,  8,
+                       9, 10, 11, 12),
+     nrow = 3, ncol = 4, byrow = TRUE))
     [,1] [,2] [,3] [,4]
[1,]    1    2    3    4
[2,]    5    6    7    8
[3,]    9   10   11   12
```

---

[4]An object can also be defined as a factor using the **factor** function.

Unlike data frames, all entries of a matrix have to be of the same type, typically (but not necessarily) numeric. Every entry in a matrix such as x is represented by notation of the form x_{ij} (or x[i,j] in R) where i identifies the row of interest and j the column.

An *array* can have any number of dimensions; a vector is a 1-dimensional array and a matrix is a 2-dimensional array. Going further, a 3-dimensional array with dimension lengths $3 \times 4 \times 2$ contains data entries identified by notation of the form x_{ijk} (or x[i, j, k] in R). Like matrices, the entries of an array all have to be of the same type. Section 6.2.5 provides an example of how a 3-dimensional array can be created and used.

1.6   Operators

In R, these specialized functions perform specific operations on one object (unary operators) or two objects (binary operators). This section illustrates basic applications of a selection of operators that find use in later chapters.

Arithmetic Operators

Arithmetic operators are denoted by, and used in much the same way as in a calculator. Let x and y be numeric vectors of equal length. Then,

Enter	To compute
x + y	termwise additions of the entries of y to those of x
x - y	termwise subtractions of the entries of y from those of x
x*y	termwise products of the entries of y with those of x
x/y	termwise divisions of the entries of x by those of y
x^y	termwise powers of the entries of x to those of y

As long as the operations are well-defined, in each case the output will be a numeric vector having the same length as x and y. For example, first create a couple of vectors

```
> x <- c(1, 4, 3); y <- c(2, 1/2, 2)
```

Then

```
> x + y # Addition
[1] 3.0 4.5 5.0
> x - y # Subtraction
[1] -1.0 3.5 1.0
```

```
> x*y   # Multiplication
[1] 2 2 6
> x/y   # Division
[1] 0.5 8.0 1.5
> x^y   # Exponentiation
[1] 1 2 9
```

A formula can be used to perform any combination of two or more of these operations, for example,

```
> 1 + 2*x/3 - y/7
[1] 1.380952 3.595238 2.714286
```

See Section 6.1 for more on computing with numeric vectors.

Relational Operators

At the simplest level, a *relational statement* is one that involves a comparison of two quantities using a *relational operator*. Such statements can have one of two values, TRUE or FALSE. Suppose a and b are two vectors of the same type and length, then corresponding entries of these vectors can be compared in a variety of ways.

Enter	To perform pairwise tests on entries of a and b
a == b	entries of a are equal to those of b
a < b	entries of a are less than those of b
a <= b	entries of a are less than or equal to those of b
a > b	entries of a are greater than those of b
a >= b	entries of a are greater than or equal to those of b
a != b	entries of a are not equal to those of b

In each case, the output is a logical vector having the same length as a and b. For example,

```
> a <- c(1, 2, 3, 4); b <- c(4, 2, 3, 1)
> a == b # Test for equality
[1] FALSE TRUE TRUE FALSE
> a != b # Test for inequality
[1] TRUE FALSE FALSE TRUE
> b <= a
[1] FALSE TRUE TRUE TRUE
```

When using these operators in the manner illustrated above it is important
to make sure the lengths and types of the two vectors are identical. If lengths
differ, an error message is printed to the screen. However, if the types differ,
logical values are still computed (see script file for examples), but the results
of such evaluations may not be meaningful for later tasks.

An exception to the equal-length suggestion applies when one of the vectors
is of length 1. For example,

```
> 2.5 > a
[1] TRUE TRUE FALSE FALSE
```

determines that 2.5 is greater than the first two entries of a.

Logical Operators

Logical values obtained, for example, from a collection of relational operations such as those described in the previous section can be combined using
logical operators to create *logical statements*. Suppose a and b are logical vectors of equal length. Then

Enter	And for each pair of entries get TRUE
p & q	only if entries of *both* p and q are TRUE
p | q	if entry of *at least one of* p or q is TRUE
!p	if entry of p is FALSE

For example,

```
> p <- c(TRUE, TRUE, FALSE, FALSE)
> q <- c(TRUE, FALSE, TRUE, FALSE)
> p & q    # Evaluate p and q
[1] TRUE FALSE FALSE FALSE
> p | q    # Evaluate p or q
[1] TRUE TRUE TRUE FALSE
> !p       # Evaluate not p
[1] FALSE FALSE TRUE TRUE
```

When using these operators in this manner, it is important to make sure
the lengths and types of the two vectors are identical. As with relational
and arithmetic operators, a length 1 logical vector can be combined with any
other logical vector. Examples in later chapters illustrate the use of compound
logical operations involving a combination of relational and logical operators
in performing data searches.

1.7   Functions

In R, a *function* is an object that takes in, as arguments, one or more other objects and performs a task on the objects.[5] Objects of the form `plot`, `summary` and others already seen belong to the class function. A further discussion on functions is given in Section 3.7. The core of much that is presented in this book is about making use of functions available in the default installation of R. There is an enormous collection of functions available on CRAN (see the next section), and users of R can also prepare their own functions; see Section 6.5 for an introduction, and later chapters for further examples.

1.8   R Packages

An R *package* typically includes a collection of functions designed for certain tasks, and sometimes datasets, along with relevant documentation. The base installation of R includes certain default packages, the contents of which can be used to prepare code to handle all tasks discussed in this book, and much more.

Default Packages

To see which packages are loaded in the current workspace, run

```
> (.packages())
```

Alternatively, the function call `search()` can be used. For the base installation of R, the seven packages displayed on running the previous code are loaded into the workspace on start-up.

To find out which additional packages are installed on your computer, run

```
> library()
```

or

```
> .packages(all.available = TRUE)
```

and the contents of a particular package that is installed on your computer can be listed using, for example,

```
> help(package = "stats")
```

---

[5]Arithmetic, relational, and logical operators are also functions; see illustrations in the script for this section.

or simply ?"stats". This opens an html window containing links (with accompanying brief descriptions) to R documentation pages for contents of package stats.

If a package is not installed on your computer you can perform an RSiteSearch. For example, to find the package car [56], either run

```
> RSiteSearch("car")
```

in the console, or through the RGui menu sequence **Help→search r-project.org. . .** and then type car in the "Question" window that pops up. The "R Site Search" window that opens contains a large list of documents matching the query.

Other Packages on CRAN

The *Comprehensive R Archive Network (CRAN)* is a vast resource for free R packages. These packages cover applications of R to a wide range of fields, and many are associated with statistics texts at various levels and in a variety of areas. In order to use a function (or any other object) from a package that is not contained in the base installation, the package in question first needs to be *installed* on your computer and then *loaded* into the workspace. Here is one way to do this.

For Version 3.4.4 of R, the RGui toolbar sequence **Packages→Install Package(s) . . .** opens the "Secure CRAN mirrors" pop-up menu. In this pop-up menu, select any one of the [**https**] mirrors listed (typically the one closest to you in location) then click on **OK**. Once this is done, the "Packages" window pops up. Now consider installing the packages moments [78], multcomp [71], and timeDate [142]. To do this, scroll down to moments in the "Packages" window and select it by left-clicking your mouse on the name. Now scroll down to multcomp, then press the **Ctrl** key as you left-click your mouse on the name, and so on. When you are done, click on **OK** to begin the process of installing the packages in a chosen directory on your computer.

To load a particular (non-default) package that is installed on your computer into the workspace, for example, package moments, run[6]

```
> library(moments)
```

Then running search() or (.packages()) will show that moments is contained in the workspace. To remove moments from the workspace, run

```
> detach(package:moments)
```

Then running search() or (.packages()) will show that moments is no longer contained in the workspace.

---

[6]Alternatively, the function call require(moments) can be used, and is sometimes preferred.

1.9   Probability Distributions

A quick tour of a typical R documentation page for a given probability distribution is appropriate at this point.

Package `stats` contains a large number of functions for a wide range of probability distributions, and up to four functions of interest are listed in the documentation page for the majority of the probability distributions.

For example, the functions for the normal distribution include `dnorm`, `pnorm`, `qnorm`, and `rnorm`. The first letter in each function's name identifies the task the function is designed to perform: `dnorm` computes *probability densities* corresponding to given quantiles (x-values); `pnorm` computes (*cumulative*) *probabilities* corresponding to given quantiles; `qnorm` computes *quantiles* corresponding to given probabilities; and `rnorm` generates a sample of *random deviates*. An illustration of using this last type of function is provided in Section 2.2.

The two traditional tasks involving the use of probability distributions at the elementary statistics level are computing probabilities (including p-values) and finding quantiles corresponding to given probabilities (critical values). Illustrations of using R's built-in density, probability and quantile functions for the various probability distributions that appear in this book are contained in later chapters (as needs arise) under sections titled "Relevant Probability Distributions." For example, uses of the functions `dnorm`, `pnorm` and `qnorm` are demonstrated in Section 11.1.3.

1.10   Coding Conventions

The code presented in this book and the accompanying script files follows certain conventions with respect to object names and indentations for multi-line function calls and computations. These conventions have been adapted from ideas contained in [6]. See also the page

```
http://adv-r.had.co.nz/Style.html
```

on the companion website for [139] for further ideas.

Single Word Object Names

All single-word object names are written in lower-case letters. For example, a vector containing weights of 35 pink salmon might be called `weights` and the function used to calculate the mean weight of the sample is called `mean`.

When a name comprises multiple words, two conventions mentioned in [6] are used, sometimes in combination. Be aware that not everyone uses the same convention. So, expect deviations in naming conventions when objects from various packages in CRAN are used in the code presented in this book.

Function Names

The names of all user-defined functions prepared in this book are written using the `period.separated` convention or a combination of the `period.separated` convention and the `lowerCamelCase` convention. For example, a function that is designed to prepare one or more specialized distribution tables for student examination scores might be named `score.distributions`. Examples that use a combination of the two naming conventions are `leveneType.test` and `errorBar.plot`, see Sections 6.5 and 10.5.

All Other Object Names

All other objects having multiple-word names are named using the `lowerCamelCase` convention. Such objects typically include vectors, factors, data frames, lists, matrices, and arrays. In this convention the letters for the first word are all lowercase, and additional words begin with uppercase letters.

For example, a data frame containing the relevant data for student exam scores might be named `studentScores`, and a list containing results of a collection of tests performed on these scores might be named `scoreTestResults`.

Code Indentation

Indentations are used when code for a particular function call or computational formula extends beyond a single line. When code extends to more than one line, indentations of four spaces are used for each new line (some use two spaces instead of four). For example,

```
plot(x = cars, xlab = "Speed (mph)",
    ylab = "Stopping distance (ft)",
    main = "Stopping Distance vs. Vehicle Speed")
```

In this example the new lines begin at well defined locations (after a comma) within the function call, and the same indentation is used for all additional lines.

There are occasions when nested indentations are used. Consider, for example,

```
with(data = corStudy,
    expr = cor.test(x = indepVar, y = dependVar,
        alternative = "greater", conf.level = 0.99))
```

Here the function `cor.test` is called inside the `with` function, and extends to two lines. The second indentation of four spaces in the third line is for the `cor.test` function's extra line.

There are occasions in this book when these indentation conventions may be broken for purposes of improving the clarity of the code. For example, in

```
x <- matrix(data = c(1,   2,   3,   4,
                     5,   6,   7,   8,
                     9,  10,  11,  12),
        nrow = 3, ncol = 4, btrow = TRUE)
```

the (non-standard) indentations used for lines 2 and 3 separate and align the rows of the matrix created.

Some Cautions: Care must be exercised when dealing with computational formulas that extend over two or more lines. Here is a very simple illustration of what can go wrong. This first run contains code that correctly performs a desired computation.

```
> 2.5 - qt(p = 0.05, df = 13, lower.tail = FALSE)*
+     1.2345/sqrt(14)
[1] 1.915709
```

The symbol "*" at the end of the first line with nothing after it indicates that the computation is not complete, and the interpreter appends instructions contained in the code on the next line to this line. Notice what happens if the "*" is moved to the start of the second line. That is, the following code is run.

```
2.5 - qt(p = 0.05, df = 13, lower.tail = FALSE)
    *1.2345/sqrt(14)
```

The result is

```
> 2.5 - qt(p = 0.05, df = 13, lower.tail = FALSE)
[1] 0.7290666
>     *1.2345/sqrt(14)
Error: unexpected '*' in "  *"
```

The first line of code is a complete computation, so the multiplication in the second line is unexpected. For this case, this issue can be remedied by placing the whole formula within parentheses, for example,

```
> (2.5 - qt(p = 0.05, df = 13, lower.tail = FALSE)
+     *1.2345/sqrt(14))
[1] 1.915709
```

works fine. So, the moral of the story when splitting code into two or more lines is to always leave the R interpreter hanging, expecting more when moving to the next line. In later chapters indentations are also used in code for other structures such as user-defined functions, loops and conditional statements.

1.11   Some Bookkeeping and Other Tips

While not necessary, the first two suggestions given here can be useful for long sessions with R. The last two tips may or may not find use, depending on the tasks being performed.

Tidying the Workspace and the Console

As mentioned previously, the `search` function is useful for checking which non-default packages are present in the workspace, and the `detach` function is used to remove non-default packages that are no longer needed.

The function `ls` can be used to list all user-generated objects contained in the workspace. For example, to find out which objects are present in the workspace at this point, run

```
> ls()
```

To remove a selection of user-generated objects from the workspace, run, for example,

```
> rm(a, b, c)
```

and to remove all user-generated objects contained in the workspace, run

```
> rm(list = ls(all = TRUE))
```

Finally, the console can be cleared by using the keystroke sequence **Crtl+L**.[7]

Changing the Working Directory

While not necessary, at the start of any R session it is convenient to set the current (working) directory. For example, while working through this chapter the working directory was set to

```
Z:\Docs\AppStats\ScriptFiles\Chapter1
```

using the toolbar selection **File→Change Dir...**, and then selecting the desired directory. Alternatively, the `setwd` function can be used, for example,

---

[7]The keystroke sequence for Macintosh users is the same.

```
> setwd(dir = "Z:/Docs/AppStats/ScriptFiles/Chapter1")
```

This code works for all operating systems. Alternative Windows specific code for this is

```
> setwd(dir = "Z:\\Docs\\AppStats\\ScriptFiles\\Chapter1")
```

Reformatting Numbers

Numbers can be truncated using the `trunc` or `as.integer` functions. For example, if

```
> x <- c(-1.5731, 0.00397, 1.2552, 2.0172, 2.9927)
```

then the output to the console from running `as.integer(x)` and `trunc(x)` are the same. The `floor` function rounds numbers down to the nearest integer, and `ceiling` function rounds numbers up to the nearest integer, for example,

```
> floor(x); > ceiling(x)
```

The `round` function rounds numbers (up or down) to a specified number of decimal places, for example,

```
> round(x, digits = 3)
```

and the `signif` function can be used to return numbers to a specified number of significant digits, for example,

```
> signif(x, digits = 3)
```

The number of digits to which numbers are printed to the console can also be altered without actually changing the value of the number stored in the workspace.

Altering the Display of Numbers on the Console

The output from running

```
> options("digits")
```

indicates that, by default, the number of digits in numeric values that are printed to the console are at most 7. See, for example, the output produced by

```
> (x <- pi); (y <- 1234567898765); (z <- 0.000000345678)
```

To change this setting to 3 digits and view the results, run

```
> options(digits = 3); x; y; z
```

or to change the setting to 15 digits and view the results, run

```
> options(digits = 15); x; y; z
```

Observe that the numbers themselves are not changed, only the manner in which they are printed to the console. To reset the digits option back to the default setting, simply run `options(digits = 7)`.

1.12   Getting Quick Coding Help

In addition to the menu sequence **Help→HTML help...** addressing contents of packages loaded in the workspace, and **Help→search.r-
project.org...** (the RSiteSearch function), topic-specific help and keyword
searches can be performed. Also available are examples of function uses and
demonstrations of a variety of features.

For example, running ?"plot" opens the *R Documentation page* for the
function plot. This can also be performed using

```
> help(topic = "plot", package = "graphics")
```

While enclosing the topic within quotes (for both approaches) is not always
necessary, it is a good idea to remember their use. For example, while ?plot
works, ?+ and ?function will not.

If a package is installed on your computer, but not loaded into the
workspace then use the help function along with the arguments topic and/or
package to get at what you want. The package argument assignment for topic
searches is not needed if the package in question is loaded in the workspace.

If the actual function name is not known one may perform keyword
searches of topics. For example, running ??"analysis of variance" performs a keyword search of topics in the area of analysis of variance. In this
case a page containing links to candidate documentation pages opens. This
search can also be initiated using

```
> help.search(pattern = "analysis of variance")
```

Always use quotes for keyword searches when more than one keyword is used.

Once the document page name of a particular "topic" has been identified
the example function can sometimes be used to run sample code provided in
the help pages. For example, running

```
> example(topic = plot, package = "graphics")
```

in the console and then following prompts in the console runs the example code given in the documentation pages for the plot function. Including
the argument assignment package = "graphics" is not necessary for default
packages, or if the package in question has been loaded in the workspace using
the library or require functions.

Another function that is useful for exploring some features in R is the
demo function. Running the function call demo() produces a list of available
demonstrations. As an example, run

```
> demo(topic = colors, package = "grDevices")
```

then follow the directions that appear on the console.

2

Bringing Data Into and Out of R

This chapter provides introductions to some basic ways to create and save data structures in R, ways of importing data into an R session, and approaches to exporting data from an R session. The methods illustrated can be extended to larger datasets. For a more complete and technical discussion of the import/export capabilities of R the reader is referred to [113].

2.1   Entering Data Through Coding

While the following can be done on the console, a convenient approach is to open a script file, enter code to create the desired data structure in the script file, check for typographical errors, and then run the code to place the resulting data structure in the workspace.

Creating Vectors

Open the script file for this chapter, `Chapter2Script.R`, and then highlight and run the following lines of code contained in the file in the manner described in Section 1.2. This code creates two generic samples, stored in the vectors a and b.

```
> a <- c(19, 53, 42, 17, 15, 27, 19, 34, 42, 48,
+        46, 53, 2, 41, 19, 31, 27, 26, 22, 32)
> b <- c(35, 69, 16, 49, 42, 21, 42, 28, 67, 39, 28, 73, 16)
```

To list the objects placed in the workspace, run `ls()`, and the contents of these two objects can be displayed by entering a and b in the console. The above code can also be saved in a separate script file, see `Data02x01.R`, and sourced when the data are needed.

For example, by selecting **File→Source R code...** from the RGui toolbar, and then selecting `Data02x01.R` from the appropriate folder, a function call of the form

```
> source("Z:\\Docs\\AppStats\\Chapter2\\Data\\Data02x01.R")
```

appears on the console, running the code contained in `Data02x01.R` just as if it were entered on the console, or run in an open script file.[1]

Alternatively, this task can be accomplished by running

```
> source(file = file.choose())
```

and then navigating the "Select file" window to select the file `Data02x01.R`.[2]

Just as it was used to combine a collection of data values, the combine function, c, can be used to combine two or more vectors. For example,

```
> (observations <- c(a, b))
```

combines the vectors a and b into a single vector, the order of the entries being unaltered and the entries of b follow those of a.

Defining Factors

Categorical data can be stored in a character vector in much the same manner as numerical data are stored in a numeric vector. For example, one can create a character vector containing class labels for the data in `observations` as follows.

```
> (groups <- rep(x = c("A", "B"),
+       times = c(length(a), length(b))))
```

Notice the convenience of using the `rep` function to generate the desired number of A's and B's.[3] Also, notice that the output shows the A's and B's enclosed in quotes, indicating these data are stored as characters.

If any form of analysis on categorical data is to be performed the vector should be converted into a factor. Turning any vector into a factor simply requires redefining it as a factor. For example,

```
> (groups <- as.factor(groups))
```

---

[1] Alternative code for the source function call is, for example,

```
source("Z:/Docs/AppStats/Chapter2/Data/Data02x01.R")
```

This form (using "/" instead of "\\") works for all operating systems, and is needed for Unix and Linux, and Macintosh operating systems.

[2] There are occasions when the function `file.choose` gives rise to problems: the computer appears to freeze up when `file.choose` is called within other functions (for example, source, save, load, among others). This is particularly the case when R is being run on the Macintosh operating system. Should this problem be encountered, the solution is to assign file names and paths to the `file` argument, for example,

```
file = "theFileLocation/theFileName.ext"
```

[3] See Section 2.2 for further illustrations of using the `rep` function.

converts `groups` into a factor that has two *levels*. The contents of `groups` are now of type integer (rather than character), each integer entry being assigned a label (such as `A` or `B`). The function `factor` can be used in place of the function `as.factor` to produce the same results. Another way to create a factor is to use the `gl` function, which makes use of the `rep` function to create factors having levels with the same number of replicates.

Building Data Frames

The vector `observations` and the factor `groups` can be placed in a data frame. For example,

```
> biVarSamp <- data.frame(observations, groups)
```

creates the data frame `biVarSamp` containing the columns `observations` and `groups`. In this case the column names in the data frame are inherited from the objects included.[4] A desired number of leading rows can be looked at using the `head` function, for example,

```
> head(x = biVarSamp, n = 4)
```

lists the first four rows in `biVarSamp`. Similarly, the `tail` function can be used to look at a desired number of rows at the tail-end of the data frame.

A compact display of the internal structure of an object, such as a data frame, can be obtained using the `str` function. For example, running

```
> str(object = biVarSamp)
```

identifies the class of `biVarSamp`, gives the number of observations and variables contained in this data frame, and also provides information on each column contained in it. The details permit a quick, brief view of the contents of a data frame.

Lists

A list can be used to store a wide variety of R objects (not necessarily "traditional" data structures). For purposes of this section its use in storing two or more vectors and/or factors having different lengths is illustrated.

For example, the vectors `a` and `b` can be placed in a list object as two separate vectors, `sample1` and `sample2` using

```
> (twoSamps <- list(sample1 = a, sample2 = b))
```

---

[4]To give the columns different names in the construction stage use, for example,

```
> data.frame(obs = observations, grps = groups)
```

Unlike data frames, the names of the objects placed in a list are not inherited, hence the assignments `sample1 = a` and `sample2 = b`.

As with data frames, the names of the objects in a list can be displayed using the `names` function and the function `str` can be used to get a feel for the internal structure of `twoSamps`.

Matrices

There are occasions when a *matrix* can be useful for storing data and displaying results, and for computations in more advanced settings. Like a data frame, a matrix is a 2-dimensional structure, but unlike a data frame all elements of a matrix have to be of the same type. That is, a matrix is either (entirely) of type numeric, or of type character, and so on.

The usage definition of the `matrix` function along with arguments and their default settings is

```
matrix(data = NA, nrow = 1, ncol = 1, byrow = FALSE,
    dimnames = NULL)
```

Here is an example in which the use of all arguments is illustrated. The matrix created here contains (as characters) the indices for each entry, and each row and column is named. By setting `byrow = TRUE`, the matrix is filled by rows (as opposed to by columns).

First use the `paste` function to generate character vectors that contain the row and column names.[5]

```
> rowNames <- paste("row", 1:3, sep = "")
> colNames <- paste("col", 1:4, sep = "")
```

Next, using the `paste` function, create a vector that contains the indices for each entry (arranged for insertion by rows).

```
> entry <- paste(rep(x = 1:3, each = 4,),
+                 rep(x = 1:4, times = 3), sep =",")
```

Now populate and view the matrix

```
> (matrixDemo <- matrix(data = entry, nrow = 3,
+     ncol = 4, byrow = TRUE,
+     dimnames = list(rowNames, colNames)))
```

Then, for example, to access the entry contained in the second row and third column, run

---

[5]See Section 7.6.3 for further details and examples concerning the `paste` function, and Section 2.2 for more about the colon, ":", operator.

```
> matrixDemo[2, 3]
[1] "2,3"
```

Traditionally, the units of measurement of the data in a matrix are all the same; however, this is not necessary as long as care is taken when using matrices having rows (or columns) with different units.

2.2 Number and Sample Generating Tricks

Quick access to data is helpful for testing code, and R has some very useful functions for such purposes. This section revisits the various number generating functions already seen, and introduces some others.

The Colon Operator and Number Sequences

As was seen earlier, for any two integers a and b, the code `a:b` produces a sequence of consecutive integers starting at a and ending at b. Further examples include, for example,

```
> 7:1; 1.5:5.5; (1/3):(5/3)
```

These show that the numbers a and b need not be integers, and the relative values of a and b determine if the resulting sequence is increasing or decreasing. Observe also that while the first term in the sequence always equals a, the sequence will stop short of b if b does not have the form $a + n$ (or $a - n$ for decreasing sequences), n being a positive integer.

The colon operator can be used in a variety of ways. For example, a geometric sequence can be generated using

```
> 2*(1/2)^(0:5)
```

Or, the powers of a sequence of integers can be obtained

```
> (1:5)^3
```

In general, for any function $f(x)$, if x is replaced by a sequence of numbers, $a : b$, then a new sequence is created using the computational rule defined by f on the elements of the sequence $a : b$.

The seq Function and Regular Sequences

The function `seq` can be used to generate a variety of regular sequences. For example, an arithmetic sequence of a desired length starting from a given value a up to at most a given value b can be obtained using,

```
> seq(from = 1, to = 3, length.out = 5)
```

Equivalently, instead of specifying the length of the sequence, an increment can be defined

```
> seq(from = 1, to = 3, by = 0.05)
```

As with the colon operator, decreasing sequences can be obtained by switching the roles of a and b. For example,

```
> seq(from = 3, to = 1, length.out = 5)
> seq(from = 3, to = 1, by = -0.05)
```

Just as for the colon operator, it is possible to use the seq function to create a variety of types of sequences.

Replications and the rep Function

An earlier application of this function took on the form

```
> c(rep(x = "A", times = 2), rep(x = "B", times = 3))
```

More compact code to perform the same task is

```
> rep(x = c("A","B"), times = c(2,3))
```

Now try

```
> rep(x = c("A","B"), times = 2)
> rep(x = c("A","B"), each = 3)
> rep(x = c("A","B"), each = 3, times = 2)
```

The input, x, can be a vector of any type and length. While the argument times may be a positive integer or a vector having the same length as x, the argument each can only be assigned positive integer values. Character vectors created as above can then be converted into factors if so desired.

Random Samples and the sample Function

A simple example of using this function is

```
> numbers <- 1:25
> sample(x = numbers, size = 10, replace = TRUE)
```

If replace = FALSE is used, then the random selection is done without replacement. Further runs of

```
> sample(x = numbers, size = 10, replace = TRUE)
```

will produce different samples. Since the function `sample` actually generates a *pseudo-random* sample selected from the sample space assigned to x, it is possible to duplicate a random sample by setting the *seed* for the *random number generator* to a chosen value each time the `sample` function call is made. For example, any number of runs of the two lines

```
> set.seed(seed = 5)
> sample(x = numbers, size = 10, replace = TRUE)
```

produce the same random sample. The `seed` can be any positive integer.

Another capability of the function `sample` is that probabilities can be assigned to elements of a set to be sampled. For example, first define a *sample space* and a list of probabilities for each element of the sample space.

```
> fruit <- c("Apples", "Bananas", "Kiwis", "Grapes")
> probs <- c(1/8, 3/8, 3/8, 1/8)
```

The probability vector must have the same length as the sample space vector and its entries must sum to 1. Then a random sample of size 3 with replacement can be simulated using

```
> sample(x = fruit, size = 3, replace = TRUE, prob = probs)
```

If the random number generator seed is not set to the same number each time this code is run, a different sample is (typically) obtained.

Samples of Random Deviates

Suppose a sample of size 23 is desired such that it simulates a scenario in which the underlying random variable is normally distributed with mean $\mu = 11$ and standard deviation $\sigma = 1.2$. Then,

```
> rnorm(n = 23, mean = 11, sd = 1.2)
```

does the job. To duplicate a sample, the `set.seed` function can be used prior to each `rnorm` function call. For example, repeated runs of

```
> set.seed(seed = 3); rnorm(n = 23, mean = 11, sd = 1.2)
```

will produce the same sample.

Functions to generate samples of random deviates for variety of probability distributions are available in R, for example, uniform (`runif`), binomial (`rbinom`), and several others that are contained in package `stats` and package `mvtnorm` [61].

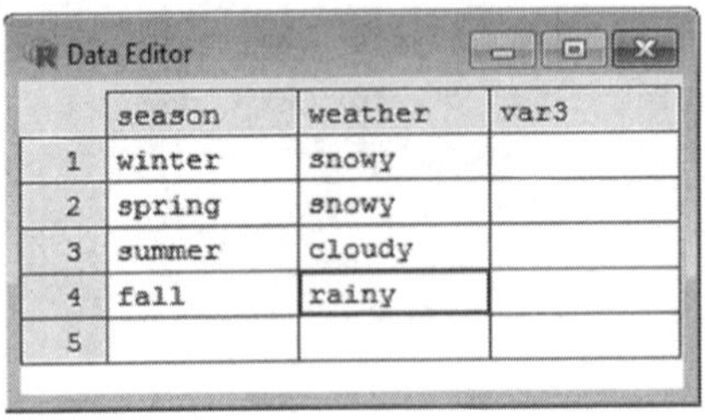

FIGURE 2.1
The R Data Editor window can be opened to perform data entry, or changes, using the `edit` or `fix` functions.

2.3 The R Data Editor

The `edit` function can be used to perform most of the tasks described in Section 2.1, but in a more organized manner.

Consider creating a small table which describes the seasonal weather in the Juneau area. First create an "empty" data frame,

```
> (juneauWeather <- data.frame())
```

then run

```
> juneauWeather <- edit(name = juneauWeather)
```

to open the data frame in the *R Data Editor*. Next, in the data editor window that pops up, click on the name space for the first column (has `var1` at this point), and enter the variable name `season`. Similarly, change the variable name `var2` to `weather`. Next, enter the data just as for a spreadsheet, see Figure 2.1 for what to enter. Close the data editor when the entries are completed. The results can then be checked by running `juneauWeather`.

The `edit` function can also be used to make changes to a data frame, including changes to data entries and variable names, as well as adding more variables.[6] For example, suppose the weather description for `spring` needs to be changed from `snowy` to `slushy`. Then,

```
> juneauWeather <- edit(name = juneauWeather)
```

---

[6]The same tasks described here can be performed using the `fix` function. In this case, simply run

```
fix(x = juneauWeather)
```

The assignment statement (to `juneauWeather`) is not needed with `fix` function; otherwise, the process is as for the `edit` function.

opens the data editor, permitting changes to be made (simply replace snowy by slushy in the spring row).

One point to keep in mind here is that the variables season and weather, within juneauWeather, are defined as character vectors (as opposed to as factors). If these variables are to be converted into factors, they will have to be redefined as such. One way to do this is as follows.[7]

```
> juneauWeather <- with(data = juneauWeather,
+      expr = data.frame(season = factor(season),
+                             weather = factor(weather)))
```

Changes can still be made to the contents of season and weather, within juneauWeather; however, on closing the data editor after the changes are made a warning message will be printed to the console alerting the user to the changes in the levels of the factor in question. On occasion, a surplus/unused factor level may remain after such a change is made. See the code for this section for some relevant illustrations and one approach that can be used to remove surplus factor levels.

2.4   Reading Text Files

A basic text editor, such as the R Editor, can be used to create text files of data. For example, see the file Data02x02.txt.[8] This file can be read into a data frame by running[9]

```
> read.table(file = file.choose())
```

The "Select File" window opens, the file can then be chosen as desired, and then the console reveals the data read from the text file.

There are some points to keep in mind when preparing a data text file. First consider a file such as Data02x02.txt. The first row in this file contains

---

[7]Notice the use of the factor function, in place of the as.factor function; they both serve the same purpose here. However, the factor function has more features.

Since both season and weather are originally character vectors, the use of the function factor (or as.factor) is not necessary; by default, the function data.frame converts character vectors into factors.

Alternative code to convert a particular column in a data frame into a factor is, for example,

```
juneauWeather$season <- factor(juneauWeather$season)
```

[8]To open this file in the R Editor, use the sequence **File→Open script ...**, then locate the file under the file type "All files (*.*)" and open the file Data02x02.txt.

[9]Mac users, recall the earlier caution concerning using the file.choose function in the OS X environment, see Footnote 2 in Section 2.1

the variable names, and there is one less entry in this row as compared to the remaining rows. Next, each row-entry is separated from its neighbor(s) by a blank space, and the first column contains the row-names as case numbers. Finally, there are no missing entries in any of the rows. In this scenario, since there is one less entry in the first row, the first row is identified (by default) as the header-row, and the first column as the row-names column. In summary, the above basic `read.table` function call suffices.

Now consider importing two variations of the file `Data02x02.txt`.

Variations in File Format

In the first scenario the first column (the row-names column) is removed, see `Data02x03.txt`. Consequently, the first row containing the variable names has the same number of entries as the remaining rows. To alert R to this, run

```
> read.table(file = file.choose(), header = TRUE)
```

and choose the file `Data02x03.txt` as described earlier.

In the second scenario the first column and the header-row are both removed, see `Data02x04.txt`. Now run

```
> read.table(file = file.choose(), header = FALSE,
+      col.names = c("y","x","g"))
```

and choose the file `Data02x04.txt`.

It should be the case that all output to the console produced so far are identical. Any one of the above data frames can be assigned a name through the use of the assignment operator. For example, running

```
> set1 <- read.table(file = file.choose(), header = TRUE)
```

and then selecting the file `Data02x03.txt` results in the data frame `set1` being created.

Handling Missing Data

When it comes to missing data, the default setting for the `read.table` function in identifying such cases is to read and flag all entries containing an NA as being missing, see `Data02x05.txt`. Run

```
> set2 <- read.table(file = file.choose())
```

and choose the file `Data02x05.txt`. On looking at the contents of the data frame `set2`, it will be noticed that the missing data are identified by an NA.

One can also check each entry of a dataset for the presence of an NA "value" using

```
> is.na(set2)
```

Entries that are not available (contain a value of NA) are flagged by a TRUE. Alternatively, the columns and row (or case) numbers containing missing data can be found by running

```
> which(x = is.na(set2), arr.ind = TRUE)
```

Any letter, symbol, or even a blank space can be used to denote a missing value in a text file, and R can be instructed to identify such cases as being NA. For example, if the NA identifiers in Data02x05.txt are replaced by asterisks, as in the file Data02x06.txt, run

```
> read.table(file = file.choose(), na.strings = "*")
```

and select the file Data02x06.txt. The resulting screen output is identical to the previously loaded data frame.

If NA entries are present in a dataset, the data can be left as is and missing cases can be handled later if and when the necessity arises. Alternatively, all NA cases can be omitted using the na.omit function as illustrated in Section 2.7.

Text Files in Table Format from the Internet

If the data are contained in a text file the read.table function can be used.[10] Here is the usage definition of this function with some additional arguments that might find use for such applications.

```
read.table(file, header = FALSE, sep = "", row.names,
      col.names, na.strings = "NA", nrows = -1, skip = 0)
```

The purposes of the arguments header, row.names, col.names and na.strings are as before, and the argument file is assigned a URL (as opposed to a local file path) in the manner illustrated below. The sep argument can be left in its default setting (entries separated by white-space). It can also be assigned "," (comma-separated entries) or "\t" (tab-delimited entries; also covered by the default setting). Finally, the arguments nrows (the maximum numbers of rows that should be read) and skip (the number of rows that should be skipped *before* beginning to read the data) can also find use. It is helpful to view the data before reading it to determine the correct argument assignments.

Consider the dataset nzrivers.txt of the length of rivers in the South Island of New Zealand, contained in the Australasian Data and Story Library (OzDASL) [124]. This is a tab-delimited text file with column headers and without a row-name first column. Viewing the data online indicates there are no missing entries, no lines at the start of the file should be skipped, and all rows should be read. So one can run (output not shown)

---

[10]Alternatively, if the data are contained in a csv file, the read.csv function can be similarly used, see Section 2.5.

```
> where <- "http://www.statsci.org/data/oz/nzrivers.txt"
> (nzRivers <- read.table(file=url(where), header = TRUE))
```

to read these data, place them in the data frame nzRivers and print the data to the console.

The same task can also be performed using the scan function (see the script file). However, the read.table function should be preferred for nicely formatted data files such as nzrivers.txt.

Text Files Not in Table Format

One way to read data from text files that are not organized in table format, including lists of datasets, is to use the scan function. For such applications the usage definition of this function, with arguments of interest, is

```
scan(file, what, sep = "", skip = 0)
```

where the field delimiter argument sep can also be assigned "\t" for tabs, "," for commas, or "\n" for new lines. In its default setting, sep identifies field delimiters by white-space. The argument what defines the type of object to which the data are read and skip indicates the number of lines to skip before reading in data.

Consider, for example, the dataset concerning the position of houses in a Japanese farming village contained in the file matui.txt on the OzDASL website. These univariate data are tab-delimited rows of numbers. Since the data are not in table form, the read.table function will not work well, but the scan function takes care of matters quite nicely.

To get things started, first identify the file path.

```
> where <- "http://www.statsci.org/data/general/matui.txt"
```

Then (output not shown)

```
> (matui <- scan(file = where, what = integer()))
```

reads the data into the integer vector matui.

The readLine function can be used for the same purpose, but not quite so elegantly. The usage definition of this function, with the only argument of interest, is

```
readLines(con)
```

The connection argument con can be assigned file.choose() for local files or, for purposes of this illustration, a URL.

In using the readLines function to read data contained in files such as matui.txt a couple of steps are involved. First, run (output not shown)

```
> (matui <- readLines(con = where))
```

Observe that each line is read as a single (long) character string, producing a vector of 30 character strings. The numbers within each character string are separated by a blank space, and some exploratory code (see the script file) leads to the following code which reads the data into an integer vector.

```
> matui <- as.integer(unlist(x = strsplit(x = matui,
+            split = " "), use.names = FALSE))
```

The argument `split` in the function `strsplit` can also be assigned `"\t"` for tabs, `","` for commas, or `"\n"` for new lines.

2.5   Reading Data from Other File Formats

There will be times when data need to be imported from non-text file formats. Here are four scenarios.

A Brute-Force Approach for PDF Files

When dealing with pdf files containing small to not too large datasets the following indirect approach works fairly efficiently. In a nutshell, the first step is to convert the pdf file into a text file, the next to clean the text file (a spreadsheet program works well) so that the contents have one of the formats described in Section 2.4. Then, use the `read.table` function as described in Section 2.4 to read the data.[11] As an illustration, consider extracting daily minimum and maximum temperatures, and precipitation observed at the Juneau International Airport for January 2014. Such data can be exported to a pdf file (see, for example, the file `Climate Data Output.pdf` contained in the script/data folder for this chapter) and then Adobe Reader can be used to save this pdf file to a text file (see `Climate Data Output.txt`) which can be viewed for completeness and cleaned up as appropriate.

Alternatively, Adobe Acrobat Pro can be used to save the file `Climate Data Output.pdf` as a spreadsheet (see, `Climate Data Output.xlsx`) which can be tidied up and saved as a text file. Whichever route is taken, the contents of the end result might have the appearance of the file `JunJanRain2014.txt`.

Reading CSV Files

Files containing tables of data in *comma-separated-values* format (*csv*), see `Data02x07.csv`, have data values separated by commas. Missing data are

---

[11]For files containing large datasets the text mining capabilities of package `tm` [51] might be preferred for importing data directly into R. For such extractions the reader is referred to [50] and [52]. Another package that might find use, particularly for very large datasets, is package `ff` [1].

represented by the absence of data. Row and column names are handled in the same manner as for the `read.table` function, and missing data are flagged using the argument assignment `na.strings = ""`. For example, run

```
> read.csv(file = file.choose(), na.strings = "")
```

and select the file `Data02x07.csv` to get the same data frame as obtained earlier.

It is worth noting that if the argument assignment `na.strings = ""` is left out of the `read.csv` function call, missing data in factors will not be flagged with an `NA`. In fact, such entries are read as "empty characters" (as `""`), and are defined as one of the levels of the factor in question.[12] See the script file for illustrations of this; this issue does not arise if a numeric column has an entry that is missing.

As with text files and the `read.table` function, other symbols can be used to identify missing entries and the `na.strings` argument can be used to identify the appropriate symbol within a `read.csv` function call.

Data Formatted for Other Statistical Programs

Package `foreign` [116] contains functions to import data stored by a variety of statistical programs including, among others, Minitab, SAS, SPSS and dBase. So, for example, after loading package `foreign` in the workspace, run the code

```
> read.spss(file = file.choose(),
+     use.value.labels = TRUE, to.data.frame = TRUE,
+     use.missings = to.data.frame)
```

and select the SPSS datafile `JunJanRain2014.sav` from the folder for this chapter. Then, the data from this (SPSS) file are read to a data frame in which value labels for variables (from SPSS) are converted into factor levels and missing values are set to `NA`. While a warning message is printed to the screen, the data are read correctly. As indicated in the documentation page, this function was originally written in 2000, so changes in SPSS formats may not be fully supported.

Spreadsheets

Spreadsheet programs can be used to prepare and save data in text or csv format. Using Excel, for example, open a new workbook and first remove all but one worksheet by right-clicking the mouse on each extra sheet (for example, Sheet2) and selecting **Delete**. Next, place the variable names in the first row of the remaining worksheet (for example, Sheet1) and corresponding

---

[12]Both the `read.table` and `read.csv` functions import columns of non-numeric data as factors.

data below each variable name. Missing data can be identified, for example, by an empty space (for csv format), or some other character such as an asterisk (for text format). Finally, use the **File→Save As** option in Excel to save the file as a "`*.txt`" or "`*.csv`" type file.

For text files, if a first column of row-names is not included and missing data are identified by an asterisk, the `read.table` function call should have the appearance

```
read.table(file = file.choose(), na.strings = "*",
    header = TRUE)
```

For a csv file, if a column of row-names is not included and missing data are identified by the absence of data, the `read.csv` function call should have the appearance

```
read.csv(file = file.choose(), na.strings = "",
    header = TRUE)
```

As described previously, the imported data can be placed in a named data frame.

2.6   Reading Data from the Keyboard

Before beginning the following example, clear the workspace of all objects except `set2`, and also clear the console.

Creating a Vector from Keyboard Entries

For a preliminary illustration, consider creating a character vector containing 3 entries by reading these from the keyboard using the `scan` function. Run the code

```
peopleNames <- scan(file = "", what = "", n = 3)
```

and then click on the console to make it the active window. The following lines show up on the console (after the names are entered one by one on the lines labeled `1:`, `2:` and `3:` on the console).

```
> peopleNames <- scan(file = "", what = "", n = 3)
1: Jack
2: Jill
3: Jen
Read 3 items
```

The workspace can be checked using `ls()`, and the contents of `peopleNames` can be examined in the usual manner. The argument assignment `file = ""` instructs the `scan` function to get the data from the keyboard, the assignment `what = ""` indicates that characters are to be read, and `n = 3` specifies the number of entries that are to be read.

Creating a Data Frame from Keyboard Entries

One can create a data frame in the same manner. Suppose a data frame containing the ages and weights of three individuals is desired. First run the `scan` function *within* a `data.frame` function call, and then enter each line of data: first the individual's name, then the age, then the weight as shown below. Remember to go to (activate) the console to enter the data.

```
> peopleInfo <- data.frame(scan(file = "",
      what = list(name = "", age = numeric(),
                weight = numeric()), nmax = 3))
1: Jack 25 153
2: Jill 27 115
3: Jen 31 127
Read 3 records
```

The argument assignment

```
what = list(name = "", age = numeric(), weight = numeric())
```

identifies the input as being a list in which each record has three entries, the first being of type character and the remaining two of type numeric. The assignment `nmax = 3` specifies the number of records, or cases. The contents of `peopleInfo` are as entered.

With respect to keyboard entries, the `scan` function is probably more useful in a programming setting where user input may be desired to determine how to proceed in a particular set of tasks. See the script file for such an example.

2.7   Saving and Exporting Data

For the following illustrations, consider a "cleaned" version of the previously obtained data frame `Set2`. Remove all rows containing an `NA`, and name the resulting data frame `cleanSet` using (output not shown)

```
> (cleanSet <- na.omit(object = set2))
```

It will be observed that there are seven rows; however, for larger data frames the number of rows can be found more conveniently using the `nrow` function. Similarly, the number of columns in a data frame can be found using the function `ncol`. Alternatively, the dimensions (number of rows and number of columns) of a data frame can be found using the function `dim`.

Suppose it is preferably to rename the rows with new case numbers, say 1 through 7. Then,

```
> row.names(cleanSet) <- 1:7
```

does the job.

Four alternatives to saving data structures are illustrated below, the choice of which approach to use being dependent on individual preference or need.

As *.RData Files

To write an "external" representation of the data frame `cleanSet` to a new file named, for example, `cleanSet.RData`, run[13]

```
> save("cleanSet", file = file.choose())
```

After moving to an appropriate directory and entering the file name `cleanSet.RData` in the "Select file" window, and then pressing the "Open" button a popup window appears asking whether to create the file. Click on "Yes" to complete the process. The file should appear in the directory chosen.

Now, clear the workspace of all objects and then run

```
> load(file = file.choose())
```

At the prompt in the "Select file" window, find and choose the file `cleanSet.RData`. Then, to display the contents, run `cleanSet`. This loading task can also be accomplished using the RGui toolbar sequence **File→Load Workspace...**

The `save` function can be used to export more than one object to a `*.RData` file. For example, running code of the form

```
save("object1", "object2", file = file.choose())
```

or

```
save(list = c("object1", "object2"), file = file.choose())
```

can be used to save the objects `object1` and `object2` to the same `*.RData` file.

As Script Files

To write script code representations of the data frame `cleanSet` to a script file, say `cleanSet.R`, run

---

[13]Mac users, recall the earlier caution about using `file.choose` in the OS X environment.

```
> dump(list = "cleanSet", file = file.choose())
```

and create the new file `cleanSet.R` in which to place the data frame. Next, clear the workspace and use the RGui toolbar sequence **File→Source R code...**, or run the code

```
> source(file = file.choose())
```

to reload the data frame.

Unlike the `save` function, the `dump` function creates a script file of R code. To see the contents of this file and how the data frame is coded, open the file `cleanSet.R` using **File→Open script...** and selecting the file `cleanSet.R`. Alternatively, a data frame can be created as described in Section 2.1 and then saved as a script file as described in Section 1.2.

As Text Files

To write a data frame such as `cleanSet` to a text file, run

```
> write.table(x = cleanSet, file = file.choose(),
+     row.names = FALSE)
```

then, in the "Select file" window, go to an appropriate directory and enter the new file-name, say `cleanSet.txt`. The file will appear in this directory, and can be opened and inspected using the R script editor. Note that if you use the R editor you will have to identify the files type as "All files (*.*)" in the "Open script" window in order to see the filename.

As with earlier created text files, the data in this file can be imported using the `read.table` function as described previously.

As CSV Files

To write a data frame such as `cleanSet` to a csv file, run

```
> write.csv(x = cleanSet, file = file.choose())
```

then, repeat the process as for the `write.table` function call, the only difference being that the file name has the extension "csv," for example, `cleanSet.csv`.[14] The data in this file can be imported using the `read.csv` function as follows.

```
> cleanSet <- read.csv(file = file.choose(),
+     na.strings = "", row.names = 1)
```

The argument assignment `row.names = 1` is needed to identify the first column in `cleanSet.csv` as the row names column. Again, the file `cleanSet.csv` can be viewed in the R editor.

---

[14]Note that the argument assignment `row.names = FALSE` can be included in the `write.csv` function call if the wish is to suppress row names in the resulting csv file.

3

Accessing Contents of Data Structures

Efficient access to data contained in vectors, factors, data frames and lists helps in the analysis of data. Being the simplest of data structures, data access methods for vectors and factors are dealt with first and then these are extended to data frames and lists. So as to build on the use of functions, this chapter closes with some additional notes on functions, their arguments, and alternative argument assignments. Open the script file for this chapter, `Chapter3Script.R`, to follow along with the content presented.

3.1 Extracting Data from Vectors

A *vector* contains a collection of entries, say $x_1, x_2, \ldots, x_n$, where $n \geq 1$ and where the *subscripts* (or *indices*) indicate the order in which the entries are placed in the vector. These indices provide a direct way to access the entries of a vector.

Using Indices/Case Numbers

Generate a small generic sample using

```
> set.seed(seed = 5)
> (y <- sample(x = 55:75, size = 7, replace = TRUE))
```

The indices of a vector can be thought of as representing the *case numbers* identifying the order in which data are observed. Extracting observations having specific case numbers, then, is a simple matter of providing the case numbers of interest. For example, code of the form

```
> y[3]; y[c(3, 7)]; y[3:7]
```

extracts the entries corresponding to the case numbers identified. Entries can also be flagged for exclusion from a vector by identifying the case numbers to be excluded. For example, each of the following

```
> y[-1]; y[-c(1, 3, 7)]
```

extract all *but* the entries identified. Just as the colon operator is used above, the function `seq` can also be taken advantage of. For example,

```
> y[seq(from = 1, to = 7, by = 2)]
```

extracts all entries with odd indices.

Using Entry Names

There are occasions when it may be meaningful to assign names to individual entries of a vector. When this is done, another means of gaining access to specific entries of a vector is available. The following code creates a small vector of character strings with named entries.

```
> comment <- c(Proportion = "Use the z-test",
+              Mean = "Use the t-test",
+              Variance = "Use the chi-square test")
```

Then `comment["Mean"]` and `comment[2]` produce the same output.

There is one possible issue with having each entry named, in whatever fashion. Suppose the value of `comment["Variance"]` is to be assigned to a new variable, say `whichTest`. It is simple enough to do this using

```
> (whichTest <- comment["Variance"])
                Variance
"Use the chi-square test"
```

However, observe that the name of the entry from `comment` has also been copied to (*inherited* by) the new variable. This is not always desirable, but it can be avoided by using the [[ ]] index operator instead of the [ ] index operator. For example, by using

```
> (whichTest <- comment[["Variance"]])
[1] "Use the chi-square test"
```

the problem is solved, only the entry value has been copied.[1]

Using Logical Vectors

Suppose that the entries of interest in `y` are those that are divisible by 3. The *modulus operator*, `%%`, can be used to identify those entries in `y` which when divided by 3 yield a remainder of 0,[2] and the *relational operator*, `==`, can be used to build a logical vector. For example,

---

[1]The two main differences between the index operators [ ] and [[ ]] are: [ ] can be used to access more than one entry, whereas [[ ]] cannot; and [ ] brings out the entry value as well as the entry name, whereas [[ ]] brings out only the entry value.

[2]For two integers a and b, with $b > 1$, the *modulus* $a\%\%b$ provides the smallest non-negative integer remainder when a is divided by b. The *quotient*, $a\%/\%b$, provides the integer quotient when a is divided by b.

```
> (byThrees <- (y%%3 == 0))
> y[byThrees]
```

extracts only those entries that are divisible by 3, those corresponding to a TRUE in the vector byThrees.

In much the same manner, the use of other relational operators and the resulting logical vectors provide a very effective way to quickly search data structures for entries of interest (or concern).

3.2 Conducting Data Searches in Vectors

There are a variety of ways in which searches can be conducted within a dataset, the most typical involving the use of relational operators. R also has some specialized functions that permit searches for entries having specific properties. First, run

```
> set.seed(seed = 17)
> y <- sample(x = 27:350, size = 523, replace = TRUE)
```

to generate a new and reasonably large sample for the following illustrations.

Finding Maximums and Minimums

The minimum and maximum values in y can be found using

```
> min(y); max(y)
```

and the case numbers for entries equaling the minimum or maximum values can be identified using

```
> which((y == min(y)) | y == max(y))
```

Alternatively, the *first* entry satisfying the above conditions can be listed as follows.

```
> which.min(y)
```

Running y[y == min(y)] shows there are three 27's, the first being at case 12. Similarly, running which.max(y) and y[y == max(y)] show that the (only) maximum of 350 occurs at case 301.

Finding Entries with Specific Values

The relational operators == (*is equal to*) and != (*is not equal to*) can be used to target specific entry values in a vector. To find out if 251 occurs in y, how often, and at which case number(s), use

```
> 251 %in% y
> sum(y == 251)
> which(y == 251)
```

Note that in logical vectors an entry of FALSE has an integer value of 0 and an entry of TRUE has an integer value of 1. So sum(y == 251) simply adds all the 1's to give the number of TRUE's.

Containment of a *collection of values* in a vector can also be checked, for example,

```
> c(113, 311, 431) %in% y
```

shows that, while 113 and 311 are contained in y, 431 is not.

A function call of the form which(y == c(113, 311, 431)) will not work; however, *logical operators* can be used to expand searches involving the which function. For example

```
> which((y == 113) | (y == 311))
```

identifies all cases satisfying either one of the conditions included.

The logical operator & and the relational operator != can be used in a similar manner to exclude certain cases. For example,

```
> y[(y != 113) & (y != 251) & (y != 311)]
```

excludes all entries equal to 113, or 251, or 311.[3]

For Ranges of Values

Ranges of values can be extracted using the remaining relational operators: > (*is greater than*); >= (*is greater than or equal to*); < (*is less than*); and <= (*is less than or equal to*). For example,

```
> y[y > 300]
```

extracts all entries greater than 300;

```
> y[(y <= 100) | (y >= 300)]
```

extracts those entries that are less than or equal to 100, or greater than or equal to 300; and

```
> y[((y > 100) & (y < 300)) | (y > 500)]
```

---

[3] Alternatively, the code

```
> y[!((y == 113) | (y == 251) | (y == 311))]
```

produces the same results.

extracts entries between 100 and 300, or greater than 500. The script file contains code showing how to use the `which` function to obtain the indices of all entries within these ranges of values.

By Measures of Position

It is possible to find data entries having specific alpha-numeric *ranks* within a dataset.[4] For example, to find the 111th ranked data entry, use

```
> y[rank(y, ties.method = "first") == 111]
```

and, to find its case number, use

```
> which(rank(y, ties.method = "first") == 111)
```

In a similar manner *percentile ranks* can be brought into searches. For example, all entries in the middle 50% of the data in y can be extracted using,

```
> midFifty <- quantile(x = y, probs = c(0.25, 0.75))
> y[(y >= midFifty[[1]]) & (y <= midFifty[[2]])]
```

Observe that this suggests a way to *trim* any desired percentage of observations from the tails of a dataset.[5]

So as to expand on this discussion, add a few more data values to the vector y to create a new vector z which contains some extreme values.

```
> z <- c(1, 5, y[1:100], 725,
+        y[101:200], 713, y[201:523], 618)
```

Now, to list all entries in z (and their case numbers) whose values lie more than two standard deviations away from the mean (of the entries in z) one can use

```
> z[(abs(z - mean(z))/sd(z)) > 2]
> which((abs(z - mean(z))/sd(z)) > 2)
```

Note that, *sometimes* (very, very rarely) the `which` function and some others prefer computational expressions contained in the function call to be enclosed in parentheses. While this is not the case here, it is useful to keep this in mind when encountering error messages associated with such applications.

---

[4] See Section 5.3.3 for more on measures of position and the `rank` function.

[5] Notice the use of the double-brackets for `midFifty` in the second line of code above; the function `quantile` names each entry of the vector `midFifty` with an appropriate percentage (see Section 5.3.3 for more on this function). The use of double-brackets is not needed in this case; however, there will be occasions when not using these in a logical operation *may* lead to an error message, or (possibly hidden) incorrect results. Caution is advised whenever logical operators are used on vectors having named entries.

Entries with Special Values

There are occasions when three additional types of entries may appear in vectors: `NA` (not available), `NaN` (not a number) and `Inf` (infinite number).[6] Consider, for example, the vector

```
> w <- c(2, NA, 3, NA, 5, NaN, 7, Inf, NaN, Inf)
```

The functions `is.na`, `is.nan`, and `is.infinite` find use in identifying undesirable entries, and `is.finite` works to identify meaningful entries. For example, to extract useable numbers, run

```
> w[is.finite(w)]
```

The function `complete.cases`, on the other hand identifies all entries that are numbers. For example, to extract complete cases, run

```
> w[complete.cases(w)]
```

The `which` function can be used along with any of the "is"-functions mentioned above to identify the case numbers as in previous examples; see the script file for illustrations.

3.3　Working with Factors

Generate a small generic sample of categorical data[7] using the `set.seed` function so as to be able to duplicate the random sample.

```
> set.seed(seed = 19)
> groups <- as.factor(sample(x = LETTERS[1:5], size = 13,
+     replace = TRUE))
```

The levels of this factor can be listed using the `levels` function and the number of levels contained in `groups` can be found using the `nlevels` function.

The contents of a factor can be accessed just as with vectors. For example, the third, fourth, and fifth entries can be extracted as a factor using[8]

```
> groups[3:5]
```

---

[6]See `?NumericConstants` for details on how R handles numeric constants (such as `Inf`, `NaN`, and other numeric formats).

[7]The object `LETTERS` is one of a few constants available in R. Run `?Constants` to find out about others.

[8]Observe that the object `groups[3:5]` inherits *all* of the levels from `groups`. The surplus levels can be removed as described in Section 2.3.

or as a character or integer vector using

> `as.character(groups[3:5]); as.integer(groups[3:5])`

The `A`'s and `C`'s can be extracted using

> `groups[(groups == "A") | (groups == "C")]`

and the case numbers of the `D`'s can be found using

> `which(groups == "D")`

A query for the presence of `E`'s can be made using either of the following

> `any(groups == "E")`

> `"E" %in% groups`

and the number of `E`'s can be determined using[9]

> `sum(groups == "E")`

The methods described in this and the previous section extend to exploring the contents of data frames.

3.4 Navigating Data Frames

Since each column of a data frame is a vector (or a factor), the methods introduced in the previous two sections apply once access to these vectors (or factors) is gained. Here, the focus is on methods to extract columns, as well as methods applicable to data frames as a whole are illustrated. While the following illustrations involve a small generic data frame, the methods described extend to larger data frames.

First create a data frame named, for example, `dataTable` (see the script file). Then, on entering `dataTable` in the console, the tablular nature of the contents can be observed. Additionally, the `names` function can be used to provide a list of the column names in `dataTable` and the `str` function can be used to provide a summarized view of the structure of `dataTable`.

Using Indices and Column Names

As with vectors (or factors), indices can be used to go to specific entries of a data frame. However, when working with data frames there is the matter of *row indices* (the case numbers) as well as *column indices* (corresponding to the included variables). For example, to extract all columns of rows 9 through 11 in `dataTable`, use

---

[9]Recall that the `sum` function interprets a `TRUE` as 1 and a `FALSE` as 0. See, also, the function `all` as a sometimes useful alternative to `any`.

```
> dataTable[9:11, ]
```

The absence of indices *after* the comma in [9:11,] indicates that all columns are included. Similarly, to extract all rows of columns 1 and 3, use

```
> dataTable[ , c(1, 3)]
```

The absence of indices *before* the comma in [, c(1, 3)] indicates that all rows are included. Combining the two previous extractions,

```
> dataTable[9:11, c(1, 3)]
```

extracts rows 9 through 11 of columns 1 and 3. An equivalent way to perform this last extraction is to use column-names (that is, variable names) to identify the columns of interest. For example,

```
> dataTable[9:11, c("x", "g")]
```

duplicates the previous output.

Using the $ Operator

This operator provides access to specific columns in data frames. Once access to a chosen column is gained, methods from Sections 3.1–3.3. apply directly. For example,

```
> dataTable$x[(dataTable$x > 5) & (dataTable$x < 15)]
```

extracts all values of x between 5 and 15. Similarly,

```
> dataTable$x[((dataTable$x > 5) & (dataTable$x < 15))
+     & ((dataTable$g == "C") | (dataTable$g =="F"))]
```

extracts all values of x between 5 and 15 for which g is either "C" or "F".

Using the subset Function

Suppose the wish is to extract *all* data in dataTable for which the values of x are greater than or equal to 5 and less than 15, and g is either "C" or "F". One way to do this is to use

```
> subset(x = dataTable,
+     subset = ((x >= 5) & (x < 15)) &
+             ((g == "C") | (g == "F")))
```

As the name suggests, this function extracts a *subset* of the original data frame. The select argument can be included in the subset function call to enable the extraction of specific columns. For example,

```
> subset(x = dataTable,
+     subset = ((x >= 5) & (x < 15)) &
+               ((g == "C") | (g == "F")),
+     select = c("x", "y"))
```

extracts data within the columns x and y that satisfy the conditions specified for the subset argument.

Using the with Function

The with function enables the performance of a task directly on the contents of a data frame. For example,

```
> with(data = dataTable,
+     expr = x[((x > 5) & (x < 15)) &
+               ((g == "C") | (g == "F"))])
```

duplicates the previous extraction where the $ operator is used.

Computational tasks may also be performed inside a with function call. For example, to find the 1% and 99% quantiles[10] for x in dataTable, run

```
> with(data = dataTable,
+     expr = quantile(x = x, probs = c(0.01, 0.99)))
```

The with function is useful when tasks are to be done selectively on portions of a data frame.

Using the attach Function

This function provides a mechanism that permits direct access to the contents of a data structure, such as a data frame. To illustrate what happens when this function is used, begin by running

```
> search()
```

As mentioned in Section 1.8, this lists the default R packages loaded upon starting up R. Now run

```
> attach(what = dataTable); search()
```

Notice that the data frame dataTable is now included in the list of attached packages/objects, and direct access to the contents of dataTable is now possible. For example,

```
> x[((x > 5) & (x < 15)) & ((g == "C") | (g == "F"))]
```

---

[10]See Section 5.3.3 for further details on quantiles and the quantile function.

duplicates a previous extraction.

Once work with the contents of an attached data frame has been completed, it is advisable to detach the data frame. For example, run

```
> detach(name = dataTable)
```

to detach `dataTable`. Running the `search` function again shows that `dataTable` is no longer in the list of attached objects and packages.

While there is no danger of the *actual contents* of an attached data frame being changed, it is advisable to detach an attached object once work with it has been completed. (See Section 3.7.1 for an illustration of why this can be important.)

3.5   Lists

A few of the access/extraction methods used for data frames can be applied to lists. For example, suppose two randomly selected samples of students from a particular college are selected, one group from the natural sciences and the other from the mathematics department. Suppose a special critical thinking exam is administered to these students and the scores are noted. Then the scores can be stored in a list called, for example, `scores` (see the script file for code used to generate the samples).

As with data frames, the `names` function can be used to list the names of the contents of `scores` and the structure of `scores` can be examined with the help of the `str` function. As with data frames, the contents of `scores` can be accessed using the $ operator, or any one of the `with` and `attach` functions. The `subset` function is not designed to work directly on lists.

For example,

```
> scores$mathMajors
```

extracts the vector `mathMajors`, and

```
> with(data = scores, expr = mathMajors[mathMajors >= 80])
```

extracts mathematics major scores that are greater than or equal to 80%. Further extractions are illustrated in the script file. Once access to vectors (or factors) contained in a list has been achieved, earlier seen methods for vectors (or factors) apply.

The use of indices with lists differs a bit from data frames in that rows and columns are not applicable. For a list, an index identifies the variable (or sample) of interest. For example, `scores[1]` and `scores$mathMajors` extract the same data. However, the first inherits the name `mathMajors` but the second does not. The use of double brackets prevents the variable name inheritance, for example, try `scores[[1]]`.

One difficulty with lists is that the absence of a common definition of rows and columns makes combined searches and/or extractions of the contents of lists more involved. Consequently, it is simpler to perform the necessary extractions/searches on each variable (or list entry) one at a time.

3.6   Choosing an Access/Extraction Method

Choosing between using indices, the `$` operator, and the `subset`, `with`, and `attach` functions is typically determined by the task at hand, and often by personal preference. See the script file for two sets of illustrations.

The first set of illustrations contain examples of various ways in which the vector of values of `x` between 3 and 20 from `dataTable` can be extracted. All produce an object containing the same data, but possibly as different objects types.

The second contains two approaches to extracting a subset of `dataTable` for which the values of `x` lie between 3 and 20 and the levels of `g` are either "A" or "F". Once again, while the two approaches produce the same object, preferences of which approach to use may vary.

3.7   Additional Notes

This section provides further comments on, and some explorations of the `attach` function. Also discussed are matters concerning the arguments included in any function call.

3.7.1   More About the `attach` Function

The convenience of writing shorter code afforded by the `attach` function makes it an attractive function to use when working with data frames or lists. In fact, if attention to detail is paid, issues are not likely to arise when using this function. However, if attention to detail is not paid, some irritations and/or errors can arise. Consider the following simple illustration.

Create a small data frame, and compute the sum of the entries in the first column

```
> xy <- data.frame(x = 1:5, y = 3:7)
> with(data = xy, expr = sum(x))
```

Here the sum is 15. Now, create a transformed version of the entries in the first column of the data frame and, absent-mindedly, name it x

```
> x <- with(data = xy, expr = (x - mean(x))/sd(x))
```

Next, suppose there is a need to attach xy and perform some tasks, one of which involves the use of the sum of the entries in the first column of xy. On attaching xy, the first hint of potential trouble is an alert printed to the screen.

```
> attach(xy)
The following object(s) are masked _by_ '.GlobalEnv':

  x
```

The vector x in the workspace and the vector x in the data frame xy have identical names and are in conflict; one has to be "masked" to protect its contents. Now, continuing on without thinking too deeply, find the sum of x (*as defined in* xy, *forgetting that another* x *now exists in the workspace*). Here the sum is zero.

Note that *the transformed* x *is used* in computing the sum of x. If this were not visible (printed to the screen) the mistaken sum would not be noticed and all subsequent computations making use of the sum of x would yield incorrect results.

The original x contained in xy, however, remains unchanged. To check, run

```
> detach(xy); xy$x
```

The moral of this story? Pay attention to detail.

3.7.2   About Functions and Their Arguments

Arguments play an important role in performing a function call and, at times, determine what tasks a function performs. This chapter may be considered a reasonable introduction to how functions are used, so it is helpful to know a bit more about certain aspects of functions and the objects they generate.

The following illustrations make use of the function sample. The usage definition of this function, as described in its documentation page, is

```
sample(x, size, replace = FALSE, prob = NULL)
```

In fact, a fairly complete list of arguments for many functions can be listed using code of the form

```
> args(sample)
function (x, size, replace = FALSE, prob = NULL)
```

Typically (but not always), default arguments for functions are listed in the usage definition in forms such as `replace = FALSE`, `prob = NULL`, etc. If such arguments are left out of a function call, default values are used. The remaining arguments, such as `x` and `size`, are *typically* required arguments. However, this is also not always the case, and it is always a good idea to check the documentation pages for additional capabilities that may exist for a particular function.

In this book most non-default argument assignments in a function call are named according to the defined usage of the function (as outlined in the documentation pages). For example, for each of the following lines of code

```
> set.seed(seed = 3); sample(x = 1:4, size = 3)
> set.seed(seed = 3)
> sample(x = 1:4, size = 3, replace = TRUE)
> set.seed(seed = 3)
> sample(x = 1:4, size = 3, replace = TRUE,
+     prob = c(0.1, 0.2, 0.3, 0.4))
```

the arguments in the function `sample` are named as provided in the usage definition. This is how most function calls have been performed till now and, with some exceptions (particularly where it is obvious what the argument entered represents), will continue.

It is not always necessary to name arguments in a function call. For example, the following function calls produce results that are identical to the three previous calls.

```
> set.seed(3); sample(1:4, 3)
> set.seed(3); sample(1:4, 3, TRUE)
> set.seed(3); sample(1:4, 3, TRUE, c(.1, .2, .3, .4))
```

Note that though the names have been removed, the order of arguments as provided within the usage definition of the function is retained.

Remembering the correct order is not necessary if the arguments are named, for example

```
> set.seed(3)
> sample(size = 3, x = 1:4, prob =c(.1, .2, .3, .4),
+     replace = TRUE)
```

However, order is important if the arguments are not named in the function call. For example,

```
> set.seed(3); sample(3, 1:4, c(.1, .2, .3, .4), TRUE)
Error in sample(3, 1:4, c(0.1, 0.2, 0.3, 0.4), TRUE) :
  invalid 'replace' argument
```

Two observations should be fairly clear from this discussion. If all the relevant arguments are passed into a function as named assignments, then the order in which the arguments are listed in the function call does not matter. On the other hand, if the arguments are not passed into the function as named assignments, then the order in which the arguments are passed into the function does matter.

3.7.3   Alternative Argument Assignments in Function Calls

The function `sample` also serves as an example of using alternate formats in which arguments are passed into a function. For example,

```
> set.seed(3); sample(x = 5)
```

produces a permutation of the integers 1 through 5. Alternatively,

```
> set.seed(3); sample(x = 5, replace = TRUE)
```

produces a random sample, with replacement, of size 5. Identical results to the previous two calls are produced using the following code.

```
> set.seed(3); sample(x = 1:5, replace = FALSE)
> set.seed(3); sample(x = 1:5, replace = TRUE)
```

More often than not, such (alternative) argument assignments for a function are described in the function's documentation pages. So, browsing these pages is always helpful.

4

Altering and Manipulating Data

The access/extraction techniques introduced in the previous chapter can be taken advantage of in making alterations to data in the various data structures introduced till now. Typical alterations may include changing entries, adding or removing data entries, renaming variables, redefining variable classes, and variable transformations. The manipulations covered in this chapter include sorting data, and restructuring the manner in which the data themselves are coded. Open the script file for this chapter, `Chapter4Script.R`, to follow along.

4.1 Altering Entries in Vectors

Small generic datasets are used in the following illustrations so as to permit manageable output for viewing the results of actions taken.

```
> set.seed(17)
> (numbers <- sample(x = c(35:45, 100),
+     size = 10, replace = TRUE))
> set.seed(13)
> (chars <- sample(x = c("A", "B", NA),
+     size = 8, replace = TRUE))
```

Note that some entries in the character vector `chars` are missing (that is, `NA`).

Removing or Changing Specific Entries

Suppose the second entry in `numbers` is a mistake, then this can be excluded in an extraction in the manner described earlier using

```
> numbers[-2]
```

Alternatively, if the correct value of this entry is known to be 40, then it can be changed using

```
> numbers[2] <- 40; numbers
```

In either case, knowing the index of the entry in question is essential.

Conditional searches of the forms described in Section 3.2 can be used to identify multiple case numbers having undesireable characteristics and these can then be removed or altered in this manner, if so desired.

Removing or Changing Missing Entries

Since some entries in the vector `chars` are missing, one might choose to exclude these as described in Section 2.4 using

```
> na.omit(chars)
```

The first line of output lists all available entries; the second, under `attr(,"na.action")`, lists the indices of the unavailable entries; and the third line, under `attr(,"class")`, states that the unavailable cases were omitted to get the resulting factor.[1] If the indices of the missing cases are known, such as cases 1, 5, and 8 for `chars`, then the entries of these case numbers can be changed using, for example,

```
> chars[c(1, 5, 8)] <- c("A", "C", "C"); chars
```

Any number of entries can be altered in this manner.

Inserting New Entries

New data can be added to the end of an existing vector

```
> (chars <- c(chars, "D", "D", "C"))
```

Or, new entries can be placed at the beginning of an existing vector

```
> (numbers <- c(17, 19, 21, numbers))
```

Alternatively, as suggested by

```
> c(numbers[1:5], 29, numbers[6:13])
```

new entries can be inserted at any location in an existing vector. The function `append` can be used to perform the same tasks, often more efficiently. For example, running

```
> (numbers <- append(x = numbers, values = 29,
+       after = 5))
```

duplicates the previous insertion, and updates the vector `numbers`. The `append` function can be used to insert more than one data value after a specified location.

---

[1] As mentioned in Section 3.2, another way to obtain the case numbers for the missing values is to run

```
> which(is.na(chars))
```

Another function that can find use here is `complete.cases`. This function takes in an object, such as a vector, and outputs a `TRUE` for all cases that are available, and a `FALSE` for those that are not available.

4.2   Transformations

Data can be transformed in a variety of ways. Numeric transformations typically involve converting data from one scale of measurement to another. Numeric data can be reclassified into groups, and redefined as a factor.

Numeric Transformations

There are occasions when it is necessary to transform numeric data from one scale to another. Here, a simple illustration is provided of centering and scaling the data in the vector `numbers`.

For example, the entries of `numbers` can be centered about the sample mean, and the results can be rounded to one decimal place.

```
> round(x = (numbers - mean(numbers)), digits = 1)
```

Or, the data can be centered about the sample mean and then scaled using the sample standard deviation, again rounding the results to one decimal place.[2]

```
> round(x = (numbers - mean(numbers))/sd(numbers),
+     digits = 1)
```

As demonstrated here, a transformation can be accomplished by simply entering a formula, expressed in code form.

The `switch` function can be embedded in a user-defined function called, for example, `transformer` to perform a variety of transformations (see the code for this section in the script file).[3] Then, for example,

```
> transformer(x = numbers, how = "centered")
```

centers the data about the mean, and

```
> transformer(x = numbers, how = "naturalLog")
```

returns the natural logarithms of the data.

Moving from Numeric to Categorical Data

Numeric data can be easily reclassified under a categorical variable. For example, consider identifying all entries in `numbers` that are less than the median as being "lower" values, and those that are greater than or equal to the median as being "upper" values. The `ifelse` function accomplishes this task quite nicely.

---

[2]The built-in function `scale` can can also be used to center and/or scale the entries of a vector and numeric columns of a data frame.

[3]The use of the `switch` function is discussed in Section 6.4.3. See, also, Section 6.5 for a preliminary discussion on preparing your own functions.

```
> ifelse(test = numbers < median(numbers),
+      yes = "lower", no = "upper")
```

If the argument `test` is `TRUE`, the value for `yes` is returned and if `test` is `FALSE`, then the value of `no` is returned.

Nested calls of the `ifelse` function can also be made. For example, by first obtaining the first and third quartiles in `numbers`,

```
> q <- quantile(x = numbers, probs = c(.25, .75),
+      names = FALSE)
```

and then running

```
> (ranges <- ifelse(test = numbers < q[1],
+      yes = "lower", no = ifelse(test = numbers >= q[2],
+          yes = "upper", no = "middle")))
```

the result is a character vector which classifies the entries in `numbers` as being in the \lower", \middle\ or \upper" ranges. One caution with nested `ifelse` function calls is that attention must be paid to opening and closing parentheses.

The results of the above conversions can then be converted into a factor. For example,

```
> (ranges <- as.factor(ranges))
```

turns `ranges` into a factor with levels `lower`, `middle`, and `upper`.

A more efficient way to accomplish such conversions is by way of the `cut` function. For example, the code

```
> cut(x = numbers,
+      breaks = c(min(numbers), q, max(numbers)),
+      labels = c("lower", "middle", "upper"),
+      include.lowest = TRUE, right = FALSE)
```

duplicates the results of the previous nested `ifelse` and `as.factor` function calls.

4.3   Manipulating Character Strings

Non-numeric data (such as those contained in character vectors or factors) can be parsed to produce new factors. Consider a character vector with entries having the form "a-b-c". For example, these might represent data from a field study of deer in which a represents the month a particular deer was sighted, b the gender, and c the location (out of, say, four possibilities) where the animal was sighted. The vector used for the following examples is tags, which contains character strings of the form "a-b-c". Consider creating three factors out of this vector. Either one of two functions can be used here, first the more intuitive approach.

Using the substr Function

The usage definition for this function is

```
substr(x, start, stop)
```

and the character(s) in the string(s) contained in x are identified by the starting position (through start) and the ending position (through stop).

The function substr can be used to extract the pieces within each character string in tags: The month is identified by the first two characters, gender by the fourth, and location by the sixth. The results are then placed in their own factors.

```
> month <- factor(substr(x = tags, start = 1, stop = 2))
> gender <- factor(substr(x = tags, start = 4, stop = 4))
> loc <- factor(substr(x = tags, start = 6, stop = 6))
```

the results can then be viewed using, for example,

```
> data.frame(month, gender, loc)
```

Note that even if month, gender and loc were character vectors the function data.frame would convert these into factors.

One point to note is that the object, x, that substr acts on does not need to be a character vector for the function to work. The object x can also be numeric, logical, or a factor. The result of an extraction, however, is always a character vector.

Using the strsplit Function

With a bit of creativity, the strsplit function can also be used to lead to the same results. The usage definition of this function, with arguments of interest, is

```
strsplit(x, split, fixed = TRUE)
```

where the character(s) in x are separated using the character identified through split. Setting fixed = TRUE (non-default) instructs the function to match the separating character exactly with split. For example, some exploration leads to the code

```
> data.frame(matrix(data = unlist(strsplit(x = tags,
+               split = "-", fixed = TRUE)),
+          byrow = TRUE, ncol = 3,
+          dimnames = list(1:length(tags),
+                     c("month", "gender", "loc")))))
```

which duplicates the previous result.

Recoding the Levels of a Factor

Consider changing the labels of the levels within the factors `month`, `gender` and `loc`. Starting with the simple ones, the code

```
> levels(gender) <- c("male", "female")
> levels(loc) <- c("A", "B", "C", "D")
```

does the job.[4]

Replacing the levels in `month` with abbreviated month names can be accomplished as follows. Since the levels of month have the appearance

```
> levels(month)
[1] "02" "03" "04" "05" "06" "07" "09" "10" "12"
```

extracting the levels of `month` as integers, then using these as indices in the `month.abb` constant vector in

```
> levels(month) <- month.abb[as.integer(levels(month))]
```

inserts the correct month abbreviations as level labels.

4.4   Sorting Vectors and Factors

Continuing with the altered vectors `numbers` and `chars`, there are two functions that can be used to sort their entries: the `order` function outputs a vector containing the rearranged *indices* of the original vector, and the `sort` function outputs a vector containing the rearranged *entries* of the original

---

[4]Note that the tasks described in the previous example can also be performed inside the `factor` function. For example,

```
gender <- factor(substr(x = tags, start = 4, stop = 4),
     labels = c("male", "female"))
```

accomplishes the same task, assuming `male = 1` and `female = 2`.

vector. In both cases, tied entries (and their indices) are reported in the same order as they appear in the data.

For reference purposes, name the entries of the vector `numbers` with index numbers and print this to the screen.

```
> names(numbers) <- 1:length(numbers); numbers
 1  2  3  4  5  6  7  8  9 10 11 12 13 14
17 19 21 36 40 29 40 44 39 41 37 37 44 37
```

The first row in this output contains the index of each entry of `numbers`, the second row contains the entries themselves. Seeing the numbers and their position (index) in the vector will help in visualizing the following examples.

Using the `order` Function

In the previous output line, observe that the eighth entry in `numbers` is the largest, the thirteenth is the second largest, and so on. If `numbers` is rearranged in decreasing (numeric) order by the entries, the *indices* of the original vector are rearranged as follows,

```
> (o <- order(x = numbers, decreasing = TRUE))
 [1] 8 13 10 5 7 9 11 12 14 4 6 3 2 1
```

So, the eighth entry goes first, the thirteenth goes second, and so on. A sorting of the vector `numbers` in decreasing order of entries can now be accomplished using

```
> numbers[o]
 8 13 10  5  7  9 11 12 14  4  6 3 2  1
44 44 41 40 40 39 37 37 37 36 29 21 19 17
```

Similarly, character vectors can also be rearranged. For example,

```
> chars[order(x = chars, decreasing = FALSE)]
 [1] "A" "A" "A" "A" "B" "B" "C" "C" "C" "D" "D"
```

rearranges the contents of `chars` alphabetically. The sorting of factors is accomplished similarly.[5]

Notice that, though the end results here are rearrangements of the vectors `numbers` and `chars`, the `order` function itself produces only a rearrangement of the indices of these vectors.

Using the `sort` Function

This function produces a sorted version of a given vector. For example,

---

[5] There is, however, a caveat here. The sorting is performed by the *integer value* assigned to each factor level, and not by the alphabetical order of the labels themselves. See Section 4.6.

```
> sort(x = numbers, decreasing = TRUE)
 8 13 10  5  7  9 11 12 14  4  6  3  2  1
 44 44 41 40 40 39 37 37 37 36 29 21 19 17
```

and

```
> sort(x = chars, decreasing = FALSE)
 [1] "A" "A" "A" "A" "B" "B" "C" "C" "C" "D" "D"
```

duplicate the previous results. Again, factors are sorted with the caveat mentioned in the previous footnote.

4.5 Altering Data Frames

Load the data in the file `Data04x01.RData` as instructed in the script file. The initial illustrations in this section involve the pulling together of simulated field data of Sitka Blacktail deer on four islands in Southeast Alaska over a period of four years. These data are contained in three data frames: `lowAlt`; `highAlt`; and `skies`, and are (intentionally) designed for purposes of this section.

Being large datasets, viewing all of the contents of any one of these data frames on the screen is not practical. The `dim` function can be used to determine the dimensions, the `head` or `tail` functions can be used to get a feel for the contents, and the `str` function can be used to get a feel for the structure (and contents).

For example, among the objects contained in the data frame `lowAlt` it is found that the objects a, d, and w are numeric vectors, the objects y and m are integer vectors, and the object t is a factor.

The existence, and locations of missing values can be determined using a combination of the `any`, `is.na` and `which` functions. For example, an application of the `which` function (see script file) shows there are a total of 248 rows containing missing values, all in the sixth column under w. Recall that, if so desired, the `na.omit` function can be used to remove all rows with NA entries. This measure is not taken at this point.

Renaming Columns

Continuing with the data frame `lowAlt`, at present the column names t, y, a, d, and w are uninformative, so one might consider assigning new, more informative names to these columns.

Use `tag` for t to represent an identifier for a specific observed deer; `year` for y to indicate the year it was observed; `altitude` for a being the altitude in feet at which it was seen; `inland` for d provides the distance in miles from the maximum high-tide mark on the shore; and `weight` for w being the weight in pounds of the deer observed (if available). These names are assigned in the same order in which the corresponding columns appear in the data frame using the code

```
> names(lowAlt) <- c("tag", "year", "month", "altitude",
+       "inland", "weight")
```

The new column names can be checked using the `head` or `names` functions.

Inserting New and Identically Ordered Columns

Each deer in the study has an identification tag that was attached to it when it was first observed. Moreover, each entry in the `tag` column contains three pieces of information. For example, a tag of 02-2-3 identifies the second female (males are identified by a 1) tagged on island 3. Suppose it is wished to separate this information under three different variables, and then add these identically ordered columns to the data frame.

Continuing with this thought, first, the information in each tag can be extracted as described in Section 4.3. For example, first extract the data for the new columns. For convenience, begin with

```
> tags <- lowAlt$tag
```

then create the new variables

```
> subject <- substr(x = tags, start = 1, stop = 2)
> gender <- substr(x = tags, start = 4, stop = 4)
> island <- substr(x = tags, start = 6, stop = 6)
```

Now, to bind the resulting character vectors `subject`, `gender`, and `island` to the columns in `lowAlt`, one approach is to use

```
> deerStudy <- cbind(subject, gender, island, lowAlt)
```

Since `lowAlt` is a data frame, the function `cbind` combines the included objects by columns into `deerStudy` as a data frame. Additionally, the character vectors `subject`, `gender`, and `island` are converted into factors. The function `cbind` can be replaced by `cbind.data.frame`, or simply `data.frame` to give the same results. Running

```
> head(x = deerStudy, n = 10)
```

shows that the new columns have been included, and an inspection of the classes of the columns shows that all the columns in `deerStudy` have retained their original classes.

Rearranging Columns of a Data Frame

It is a simple matter to rearrange the columns of `deerStudy`. For example,

```
> deerStudy <- deerStudy[ , c(4, 5, 6, 1, 2, 3, 7, 8, 9)]
```

rearranges the columns of deerStudy in the order tag, year, month, subject, gender, island, altitude, inland, and weight.

Changing the Class of a Column

Suppose it is wished to also define year and month as factors. To redefine year and month as factors *inside* deerStudy the following approach can be used.

```
> deerStudy$year <- as.factor(deerStudy$year)
> deerStudy$month <- as.factor(deerStudy$month)
```

It will be found that these two columns, in deerStudy, are now factors.

Now remove all objects except deerStudy, highAlt, and skies from the workspace before continuing.

Changing Factor Level Labels

Consider assigning more meaningful names to the levels in gender and island. Running

```
> levels(deerStudy$gender); levels(deerStudy$island)
```

shows that the current levels of gender are "1" and "2", and of island are "1", "2", "3", and "4".

Since, for gender, males are identified by a "1" and females by a "2", running

```
> levels(deerStudy$gender) <- c("Male", "Female")
```

changes the levels to "Male" and "Female". Similarly,

```
> levels(deerStudy$island) <- c("Admiralty", "Baranof",
+        "Chichagof", "Douglas")
```

changes the levels of island to the indicated names, in the order given.

Inserting New Rows

Consider a scenario in which additional rows of data arrive late, or additional information that may be useful for a study is acquired. There are ways to handle such scenarios directly in R (without having to rebuild a data frame from scratch).

For the following illustration, suppose some high altitude data contained in the data frame highAlt and corresponding to the same study *and* deer arrive. The leading (or ending) rows of this data frame can be viewed using the `head` or `tail` function, and it will be found that the column names for `highAlt` match exactly with those of `deerStudy` from previous illustration. Moreover, an examination of the classes of the contents of `highAlt` shows that the classes of the corresponding columns and factor levels also match those of the altered `deerStudy` data frame.

To attach the rows in `highAlt` to the *end* of the `deerStudy` data frame, run

```
> deerStudy <- rbind(deerStudy, highAlt)
```

Using the `tail` function it will be found that the updated `deerStudy` data frame now contains a total of 1111 rows. The function `rbind.data.frame` can be used to perform the same task.

Adding New Columns by Insertion Keys

The data frame `skies` contains exactly the same data as in the revised data frame `deerStudy` from the previous example under the variables `year`, `month`, `tag`, and `altitude`, as well as some additional data under the variable `weather`. However, in the `skies` data frame the rows are ordered differently. So, inserting these new columns in `deerStudy` requires a different approach from that illustrated previously.

The function `merge` can be used to insert new columns by using insertion keys to correctly arrange and insert the entries of the new columns. For example,

```
> deerStudy <- merge(x = deerStudy, y = skies,
+       by = intersect(x = names(deerStudy),
+                       y = names(skies)))
```

does the job, and arranges the data in the new data frame according to the variables used as keys. The `intersect` function is used to extract the names of columns that are common to both data frames for use in the argument `by`.[6]

With the new variable `weather` included, the updated data frame now has 10 columns (variables).

4.6   Sorting Data Frames

For the following illustrations a small random sample of rows is extracted from the previously constructed data frame `deerStudy`, and a new data frame

---

[6]This is the default setting for the function for the argument `by`. See Section 4.8 for a further discussion on the `merge` function.

with fewer columns is created. For example, so as to be able to duplicate the following results, first set a random number generator seed, then obtain a random sample of row numbers (size 10), and identify the columns desired

```
> set.seed(13)
> cases <- sample(x = 1:nrow(deerStudy), size = 10)
> columns <- c("gender", "month", "altitude", "inland")
```

Next, extract the small sample, rename the rows, and output it to the screen for future reference purposes (output not shown).

```
> (smallSample <- data.frame(deerStudy[cases, columns],
+     row.names = 1:10))
```

As with vectors, the functions `sort` and `order` can be used to rearrange individual vectors *extracted* from a data frame. However, the matter of interest here is how to sort the *entire* contents of a data frame by one or more variables.

Sorting by a Single Variable

Consider sorting the contents of `smallSample` by `gender`. This is quickly accomplished using

```
> newOrder <- with(data = smallSample,
+     expr = order(x = gender, decreasing = FALSE))
> smallSample[newOrder, ]
```

It will be noticed that the ordering of `gender` is (alphabetically) descending, whereas the desired sort was ascending. In case this is of concern, here is an explanation of what happened.

By definition, `gender` is a factor with levels `"Male"` and `"Female"`; however, the type of the entries is `"integer"`. So the value of `Male` is 1, the value of `Female` is 2, and the function `order` views `Male` as being (numerically) less than `Female` in value. To avoid outcomes like this, one can think ahead and assign levels to factors in such a way that they are both alphabetically (by labels) and numerically (by levels) arranged in the same order. For example, if females were assigned a 1 and males were assigned a 2 in the original coding of the variable `gender`, then the numerical and alphabetical ordering would agree.

Sorting by Two or More Variables

The `order` function can be used again.[7] For example, consider performing a nested sort on `smallSample` first by `gender` and then by `altitude`, both in ascending order. Then,

---

[7]The package `rgr` [58] is one among a few packages that contains a function to perform nested sorts on data frames. See the script file for sample code.

```
> newOrder <- with(data = smallSample,
+      expr = order(gender, altitude, decreasing = FALSE))
> smallSample[newOrder, ]
```

does the job. To perform a sort on smallSample first by gender in ascending order and then by altitude in descending order, just place a "-" before altitude. For example, runn

```
> newOrder <- with(data = smallSample,
+      expr = order(gender, - altitude, decreasing = FALSE))
> smallSample[newOrder, ]
```

4.7   Moving Between Lists and Data Frames

There may be occasions when working with data in (a generic) list format is more convenient than with data in data frame format, or vice versa. For the following illustration first extract a sample of 150 randomly selected rows from the data frame deerStudy under island and altitude, and place these in a data frame named biVarFormat (see code in the script for this section). Now consider extracting altitude observations for deer classified by the island on which they were observed, and then placing the results in a list of four separate samples.

Moving from Data Frames to Lists

This can be done using the split function.

```
> (sampleFormat <- with(data = bivarFormat,
+      expr = split(x = altitude, f = island)))
```

The same task can be accomplished using the list and tapply functions.[8]

Moving from Lists to Data Frames

Moving from sample format to bivariate format can be accomplished through the use of the unsplit function as follows. First, get the numbers of deer observed on each island.[9]

```
> (frequencies <- sapply(X = sampleFormat, FUN = length))
```

---

[8]See the script file, as well as Section 6.2.2 for more on the tapply function, and others like it.

[9]See Section 6.2.1 for more on the sapply function, and others like it.

Next, build a factor from the names of the samples in `sampleFormat`, with island names as its levels.

```
> (island <- as.factor(rep(x = names(sampleFormat),
+       times = frequencies))
```

Then,

```
> backAgain <- data.frame(island = island,
+       altitude = unsplit(value = sampleFormat,
                        f = island))
```

gives the desired data frame. As demonstrated in the script file, the same can be accomplished using the `unlist` function. The result can be compared to `bivarFormat` as follows.

Comparing the Contents of Two Data Frames

Since the data in the data frames `bivarFormat` and `backAgain` are very likely ordered differently, first sort the two by `island`, and then by `altitude`.

```
> bivarFormat <- with(data = bivarFormat,
+       expr = bivarFormat[order(island, altitude), ])
> backAgain <- with(data = backAgain,
+       expr = backAgain[order(island, altitude), ])
```

The function `identical` can be used to compare the contents (and attributes) of any two given objects to determine if the objects are identical. If this function is applied to the two data frames as they are currently defined,

```
> identical(x = bivarFormat, y = backAgain)
[1] FALSE
```

By applying the `attributes` function to the objects `bivarFormat` and `backAgain` it is found that the difference between the two lies in the row-names.

It is this difference in row names that triggered the `FALSE` in the previous comparison. By renaming the rows and then comparing the two,

```
> identical(
+       x = data.frame(bivarFormat, row.names = 1:150),
+       y = data.frame(backAgain, row.names = 1:150))
[1] TRUE
```

the data frames are found to be identical.

4.8   Additional Notes on the `merge` Function

The usage definition of this function for the class `data.frame`, with arguments of interest, is

```
merge(x, y, by = intersect(names(x), names(y)),
    all = FALSE)
```

In order to make effective use of the `merge` function when combining two data frames, the columns used as keys should contain enough information to bind corresponding columns in a well-defined manner. In the default setting, the function `merge` uses all common columns in the merging process. However, there will be occasions when this will not provide enough information for a well-defined merging of two data frames. The following discussion explores this matter a bit further, and also looks at what happens when two data frames with differing numbers of rows are combined using the `merge` function.

The Issue of Repeated Values in Keys

Begin with two small data frames, `xy` and `yg` (see script file). In their current form, these two data frames can be merged using

```
> merge(x = xy, y = xg)
```

However, it will be observed that something unexpected has happened. The original data frames have 5 rows each, but the merged data frame has two additional rows. The complication occurs for $x = 2$. Since `xg` and `xy` both have two cases with $x = 2$, and since the manner in which the cases from these two data frames should be combined is not well-defined, all possible combinations of the second and third rows (4 in all) are included in the combined data frame. So, if ties occur in one of the keys, there need to be additional keys that will resolve the ties.

The Issue of Not Enough Keys

For the next set of examples, first add a new numeric variable, `f`, to both data frames (see the script file). Then, using only `x` as the key, run

```
> merge(x = xy, y = xg, by = "x")
```

Now, in addition to the two extra rows, two additional columns appear in the result. Notice that `f.x` contains the entries of `f` arranged according to `x`, and `f.y` contains the entries of `f` arranged according to `y`, both with repeats of the rows containing $x = 2$.

If only `f` is used as the key, the extra rows do not appear. However, two correctly ordered duplicate columns of `x` appear (as `x.x` and `x.y`).

```
> merge(x = xy, y = xg, by = "f")
```

Since there are two common columns, x and f, both should be used as keys, that is, use

```
> merge(x = xy, y = xg,
+      by = intersect(names(xy), names(xg)))
```

to correctly merge the two data frames. Note that it is not necessary to include by = intersect(names(xy), names(xg)) in the function call since this is the default setting for the by argument.

The Issue of Unequal Numbers of Rows

Now consider removing the second and fourth rows from xg, and then merging the resulting data frame with xy.

```
> xg <- data.frame(xg[-c(2, 4), ], row.names = 1:3)
> merge(x = xy, y = xg)
```

Only the rows corresponding to rows in xg are merged.

To get all rows from both data frames included in the merged data frame, the argument assignment all = TRUE needs to be included in the function call. For example,

```
> merge(x = xy, y = xg, all = TRUE)
```

provides all the relevant data from both data frames, including entries that are not available.

5

Summaries and Statistics

The focus here is on common methods of summarizing or describing data, including frequency distributions and the various measures of characteristics for numeric data, such as central tendency, spread, position, and shape. The emphasis in this chapter is mostly on methods applicable to univariate samples. These methods can be extended to multivariate samples using the tools introduced. Use the code in `Chapter5Script.R` to follow along.

5.1   Univariate Frequency Distributions

Frequency distribution tables (*contingency tables*) can sometimes provide a meaningful summary of what numerical data look like in the distributional sense, and are the only way categorical data can be *quantitatively* summarized. This section begins with three types of frequency distribution tables typically used for univariate data.

The data used for the illustrations in this section are a slightly edited and reordered version of the data frame `deerStudy` from the previous chapter. The function that finds use here is the `table` function.[1]

Nominal Data

Begin with obtaining a frequency distribution of deer observed by islands.

```
> (freq <- with(data = deerStudy, expr = table(island)))
```

Obtaining a sorted frequency distribution is a simple matter of applying the `sort` function.

```
> sort(x = freq, decreasing = TRUE)
```

A total column can be included using the `addmargins` function,

```
> addmargins(A = freq)
```

---

[1] An alternate function that accomplishes the same tasks is the `xtabs` function.

a relative frequency distribution can be obtained by taking advantage of the `prop.table` function,

```
> round(prop.table(x = freq), 3)
```

and, finally,

```
> round(prop.table(x = freq)*100, 1)
```

produces a percent table.

Ungrouped Numeric or Ordinal Data

Consider obtaining a frequency distribution of deer observed by months. Such frequency distributions are categorical frequency distributions if the months are labeled by name, or are equivalent to ungrouped (numerical) frequency distributions if the months are coded as numbers (1, 2, ..., 12). Either way, the code from the previous example provides the desired results. For example,

```
> with(data = deerStudy, expr = table(month))
```

If the levels of `month` are coded with actual month names using the constant `month.abb`, then the names of the columns for above table will change to month names. See the script file for additional illustrations.

For ungrouped numeric data, as long as the underlying random variable is discrete and the range of the sample space is not too large, a frequency distribution analogous to a categorical frequency distribution serves the purpose. This, however, does not work for datasets associated with continuous random variables, or even discrete random variables with sample spaces having a large range.

Grouped Numeric Data

Consider, for example, the data for altitudes within `deerStudy`. In preparing a grouped frequency distribution, knowing the sample size and the range of values in `altitude` is helpful. The `nrow` and `range` functions show that there are 1111 rows, and altitudes range from 10 through 4200 feet.

Now, the task at hand is to partition the range into non-ovelapping intervals of equal widths that cover the whole range of possible values in the sample. It is recommended that the number of such intervals be no less than 5, and no more than 20.[2]

Begin by establishing the boundaries for each interval. For example, one might use the following sequence as *class boundary values*, or *break points*, to partition the data.

---

[2]This is particularly for the case if a graphical representation of the frequency distribution is desired.

```
> (bps <- seq(from = 0, to = 4200, length.out = 9))
```

These break points partition the interval $[0, 4200]$ into eight classes, each of width 525. The next task is to identify each entry in `altitude` with one (and only one) of these intervals.

The function `cut` is designed to do exactly this, it can be used to construct a factor from the numeric vector `altitude` and the levels of this factor are defined as classes formed by partitioning the data range according to the sequence of breakpoints in `bps`.

```
> altInts <- with(data = deerStudy,
+      expr = cut(x = altitude, breaks = bps,
+          include.lowest = TRUE, right = FALSE,
+          dig.lab = 5))
```

With the above code, all intervals in the resulting factor except for the last have the form $[a, b)$; this is accomplished through the argument assignments `include.lowest = TRUE` and `right = FALSE`. The argument assignment `dig.lab = 5` is used to ensure the number of digits used in the interval labels do not get too unwieldy (not really needed here).

The resulting object, `altInts`, is a factor and the `table` function can be used to obtain the frequency distribution. A variety of approaches can be used to display the frequency distribution for `altInts`.

The `table` function can be used to print the table as a row of frequencies (output not shown).

```
> (f <- table(altInts))
```

The `matrix` function can be used to display the frequencies as a matrix with named dimensions.

```
> matrix(data = f, ncol = 1,
+      dimnames = list(levels(altInts), "Freq"))
```

The `summary` function can also be used. For example, the relevant information including columns for the *class intervals*, *class midpoints*, and *class frequencies* can be placed in, and displayed as a data frame using code of the form[3]

```
> f <- summary(object = altInts, maxsum = 8)
> mp <- seq(from = 525/2, to = 4200, by = 525)
> (altFreq <- data.frame(Mid = mp, Freq = f,
+      row.names = names(f)))
```

See the script file for an alternative display method.

---

[3]The argument assignment `maxsum = 8` instructs R to list all 8 levels. The default setting is `maxsum = 7`.

5.2 Bivariate Frequency Distributions

The methods introduced in the previous section extend to bivariate frequency distributions quite easily by simply including the two variables of interest as arguments in the `table` function (or **xtabs**) call. The levels of the first variable form the rows in the resulting table, and the levels of the second variable form the columns.

Nominal and Ordinal Data

To obtain a contingency table of the genders of deer observed by islands, including row and column totals, run

```
> with(data = deerStudy,
+     expr = addmargins(A = table(gender, island)))
```

This can be converted into a table of proportions or percentages using the `prop.table` function as described earlier; see script file for code.

Similarly, the detail in previous the frequency distribution of deer observed by months can be increased by including **gender** as an argument in the `table` function there.

Grouped Numeric Data

Consider, for example, using a partitioning of the `inland` data within `deerStudy` to construct a contingency table of the distance from the shore that deer were observed on each island.

First, as described previously, a factor of intervals is obtained. After checking the range of values in `inland`, suitable breakpoints can obtained.[4]

```
> bps <- seq(from = 0, to = 4, length.out = 5)
```

Next, construct a factor of intervals,

```
> inlandInts <- with(data = deerStudy,
+     expr = cut(x = inland,
+         breaks = bps, include.lowest = TRUE,
+         right = FALSE, dig.lab = 5))
```

and, finally, the grouped frequency distribution is obtained using

```
> with(data = deerStudy,
+     expr = table(island, inlandInts))
```

---

[4]Note, four classes are used here for convenience. As mentioned previously, the recommended number of classes in grouped frequency distributions should range from 5 to 20.

A contingency table for grouped data from two numeric variables can be constructed in a similar manner. Consider obtaining a grouped contingency table of observed frequencies for `altitude` and `inland` from `deerStudy`. For example, the data in these vectors can be partitioned into two factors of intervals, `altCuts` and `inCuts` (see code in the script file). Then, as in the previous example, the `table` function produces the contingency table desired.

```
> table(altitude = alt.cuts, inland = in.cuts)
```

The assignments `altitude = altCuts` and `inland = inCuts` enable the change in the row and column variable names.

Once again, tables of proportions or percentages can be constructed using the approach described previously, and margins containing row and column totals can be tacked on with the help of the `addmargins` function.

5.3 Statistics for Univariate Samples

For convenience, copy the `altitude` data from `deerStudy` into a new vector, say `alts`. The data in this vector are used to demonstrate the various functions available in R to compute common statistics in most of the following illustrations.

5.3.1 Measures of Central Tendency

When dealing with categorical data, the *mode*, the most commonly occuring data value (factor level), is used to identify the most typical data value in a sample. Frequency distributions, described in the previous section, provide the means of obtaining this statistic. For discrete data, or if the data are organized in a grouped frequency distribution, the *modal class* may find use; however, when the underlying random variable for a sample is continuous the mode is typically not useful and other measures are preferred and/or necessary.

Consider any numeric sample denoted by $x_1, x_2, \ldots, x_n$. Here are typical measures of central tendency that appear in elementary applications.

Arithmetic Mean

The *sample mean*, $\bar{x}$, is defined by

$$\bar{x} = \frac{\sum_{i=1}^{n} x_i}{n},$$

and is computed using

```
> mean(x = alts)
```

The same formula, and function applies for a *known* population in computing the population mean μ.

The mean is unique for each sample (or population), and is not necessarily contained in the dataset for which it is computed. This measure is sensitive to extreme values in one of the tails of the data and, consequently, is most suited for fairly homogeneous, or at least symmetric data. If the data contain extreme values in one of the tails, other measures of central tendency that are insensitive to extreme values are preferred.

Trimmed Mean

When data contain extreme values, there are many who prefer to trim values from the tails of the data before computing the mean. The $100 \times p\%$ *trimmed sample mean*, $\bar{x}_{tr(p)}$, is defined by

$$\bar{x}_{tr(p)} = \frac{\sum_{i=k+1}^{n-k} x_i}{n - 2k}, \quad k = \lfloor pn \rfloor,$$

The notation $\lfloor \cdot \rfloor$ is for the *floor function*. For example, let $x_1, x_2, \ldots, x_{53}$ be a sample arranged in ascending order, then $n = 53$. Suppose a 5% trimmed mean is desired, then, $p = 0.05$, and

$$k = \lfloor pn \rfloor = \lfloor 2.65 \rfloor = 2.$$

So, the two lowest and two highest entries are removed, leaving

$$x_3, x_5, \ldots, x_{51}.$$

The mean of these entries is called the 5% trimmed mean.

In R, the `mean` function with the argument `trim` computes trimmed means. For example, the 2.5% trimmed mean of `alts` is computed using

```
> mean(x = alts, trim = 0.025)
```

The trimmed mean, which is moderately sensitive to extreme values, is an alternative to the mean when the data contain extreme values.

Weighted Mean

Suppose the elements of the sample $x_1, x_2, \ldots, x_n$, are assigned weights $w_1, w_2, \ldots, w_n$ (which may be in the form of frequencies, percentages, or proportions that add to 1). Then the *weighted sample mean*, $\bar{x}_w$, is computed using

$$\bar{x}_w = \frac{\sum_{i=1}^{n} w_i x_i}{\sum_{i=1}^{n} w_i}.$$

In R, this is computed using the function `weighted.mean`. For example, suppose an elementary statistics student ends up with an average score of 83%

on her homework assignments, 91% on her midterm exams, 87% on her term project, and 88% on her final exam. If the weights assigned to each of these scores are 15% for homework, 50% for midterms, 15% for the project, and 20% for the final, then the weighted mean is computed as follows.

```
> scores <- c(83, 91, 87, 88); weights <- c(15, 50, 15, 20)
> weighted.mean(x = scores, w = weights)
```

Note that no matter what form the weights are entered in, the calculation of the weighted mean is equivalent to

$$\bar{x}_w = \sum_{i=1}^{n} p_i \, x_i,$$

where, for each $i = 1, 2, \ldots, n$,

$$p_i = \frac{w_i}{\sum_{i=1}^{n} w_i}$$

can be thought of as the proportion of the mean that is contributed by x_i.

Weighted Mean from a Grouped Frequency Distribution

Another application of using the weighted mean is when (an approximate to) a sample mean is desired from a (grouped or ungrouped) frequency distribution of a numeric random variable. In this case, the class midpoints are used as sample values (x_i) and class frequencies are used as weights (w_i) in the formula for computing the weighted mean.

For example, consider the grouped frequency distribution for altitude that was obtained in Section 5.1, and stored in the data frame `altFreq` with column names `Class`, `Mid` and `Freq`.

If nothing else were known about the data, a rough estimate of the sample mean could be obtained from this frequency distribution using the weighted mean.

```
> with(data = altFreq,
       expr = weighted.mean(x = Mid, w = Freq))
```

As should be expected, the results of such a computation will typically differ quite a bit from the mean obtained using raw data.

Median

Suppose the sample $x_1, x_2, \ldots, x_n$ is arranged in ascending order, the *median*, $\tilde{x}$, is obtained using the rule

$$\tilde{x} = \begin{cases} x_{(n+1)/2} & \text{if } n \text{ is odd,} \\[2mm] \left[x_{n/2} + x_{n/2+1} \right] / 2 & \text{if } n \text{ is even.} \end{cases}$$

This is computed using

```
> median(alts)
```

This measure is insensitive to extreme values in a sample and is very often the preferred measure of central tendency when the data contain extreme values in one of the tails.

Median Class from Grouped Frequency Distributions

If data are arranged in a grouped frequency distribution as, for example, in the data frame `altFreq`, then the *median class* can be found by looking for the class that contains the middle data value (by numeric rank).

Consider the case of the grouped frequency distribution in `altFreq`. Since the sum of the frequencies is 1111 (an odd number), the position of the middle value is 556. A column of cumulative frequencies (see script file) shows that the class $[525, 1050)$ is the first that has a cumulative frequency that exceeds 556. So the median class is $[525, 1050)$.

Midrange

Once again, suppose the sample $x_1, x_2, \ldots, x_n$ is arranged in ascending order. The *midrange, MR*, is defined as

$$MR = [x_1 + x_n]/2,$$

and can be computed using

```
> mean(range(alts))
```

While this measure is not commonly used, it is easy to compute and is useful if the midpoint of the range of data values is desired. This measure is also sensitive to extreme values in the data.

5.3.2   Measures of Spread

Again, frequency and grouped frequency distributions can be used to get a feel for how data (categorical or numeric) are distributed across the range of possible values. In the case of numeric data there are also some common measures of spread. Once again, consider a sample of numeric data, $x_1, x_2, \ldots, x_n$.

Variance and Standard Deviation

The *sample variance, s^2*, can be computed using any one of the formulas

$$
\begin{aligned}
s^2 &= \frac{\sum_{i=1}^{n}(x_i - \bar{x})^2}{n-1} \\
&= \frac{\sum_{i=1}^{n} x^2 - n\bar{x}^2}{n-1}.
\end{aligned}
$$

The *sample standard deviation* is simply $s = \sqrt{s^2}$.

Continuing with the vector `alts` containing altitudes at which deer were observed, running

```
> var(alts); sd(alts)
```

provides the sample variance and standard deviation, respectively.[5]

Range and Interquartile Range

The *range* is the difference between the largest and smallest data values,

$$R = x_{\max} - x_{\min},$$

and the *interquartile range, IQR*, is defined as

$$IQR = Q_3 - Q_1,$$

where Q_1 is the first quartile and Q_3 is the third quartile, see Section 5.3.3. In R, these measures for `alts` are obtained using

```
> diff(range(alts))
> IQR(alts)
```

The function `range` outputs a vector containing the minimum and maximum values in `alts`, and the function `diff` subtracts the first entry from the second entry.[6]

Also available in this area is the function `mad` which computes the *median absolute deviation*, a more robust alternative to the interquartile range.

5.3.3 Measures of Position

Again, consider any numeric sample, $x_1, x_2, \ldots, x_n$, which may or may not be arranged in ascending order. Three measures of position are commonly encountered in elementary applications: the rank of an entry, the percentile rank, and the standard score. While typical tasks involve finding measures of position for given entries in a sample, there is often a need to identify (as well as estimate) data entries that correspond to a given measure of position. Both scenarios are discussed here, measures of position and their corresponding inverse operations/processes.

---

[5]It should be remembered that the functions `var` and `sd` compute s^2 and s, *not* σ^2 and σ. The population variance is given by

$$\sigma^2 = \frac{\sum_{i=1}^{n}(x_i - \mu)^2}{n} = \frac{n-1}{n}\left[\frac{\sum_{i=1}^{n}(x_i - \mu)^2}{n-1}\right].$$

So, if `x` were to represent the data for a *whole population* one could use code of the form

```
n <- length(x); var(x)*(n-1)/n
```

to compute the population variance

[6]The `diff` function can be applied to vectors of any length (2 or more) and has broader applications to time-series data in computing *lagged differences* of entries. See, for example, [29] for a coverage of using R for time series analyses.

The Rank of an Entry

The *rank* of an entry in a sample refers, literally, to the *alpha-numeric rank* of the entry in relation to the remaining entries. The ranks of all entries in the vector `alts` are obtained using (output not shown)

```
> rank(x = alts, ties.method = "first")
```

and the rank of an entry with a specific case number can be obtained using, for example,

```
> rank(x = alts, ties.method = "first")[115]
```

By assigning `ties.method = "first"`, ranks for tied entries are assigned in the order that the tied data occur. Other options for this argument include `"average"` (the default setting), `"random"`, `"max"`, and `"min"`. Preferences of which ties-method to use vary, often by applications of this measure of position.

The ranks of specific data values can be obtained using, for example,

```
> rank(x = alts, ties.method = "first")[alts == 103]
```

and the value of an entry having a specific rank can be extracted using, for example,

```
> alts[rank(x = alts, ties.method = "first") == 373]
```

Finally, the case number of an entry with a specific rank can be found using code of the form

```
> which(rank(x = alts, ties.method = "first") == 895)
```

Percentile Ranks

The *percentile rank* of a number in a sample of not necessarily distinct entries is the percent of data entries having *ranks* that are less than the *rank* of the number in question.[7]

To compute this measure, let c_a denote the number of entries in a sample of not necessarily distinct data values, $x_1, x_2, \ldots, x_n$, arranged in ascending order that have ranks less than a number a. Then, in its simplest form, the percentile rank of a is computed using

$$p_a = c_a/n \times 100.$$

In R, the function `ecdf` can be used to obtain the percentile ranks of one or more numbers with respect to a given sample.

For example, using the vector `alts`, first create the corresponding empirical cumulative frequency distribution by running

---

[7]Alternatively, the percentile rank of a number is the percent of data entries that are less than or equal (in *value*) to the number in question.

```
> Fn <- ecdf(x = alts)
```

The object `Fn` then serves as a step-function that provides percentile ranks for any collection of numbers with respect to `alts`. For example,

```
> round(Fn(c(365, 2731))*100, 1)
```

indicates that, with respect to the data in `alts`, an altitude of 365 ft has a percentile rank of 36.9%, and 2731 ft has a percentile rank of 90.1%. See the script file for further examples.

Percentiles and Quantiles

The pth *percentile*, as typically described in elementary statistics texts, for a given dataset $x_1, x_2, \ldots, x_n$ arranged in ascending order is the number q_p that corresponds to a percentile rank $100 \times p\%$, and is obtained using the rule

$$q_p = \begin{cases} x_k & \text{where } k = \lceil np \rceil \text{ and } np \text{ } is \text{ } not \text{ an integer,} \\ \left(x_k + x_{k+1}\right)/2 & \text{where } k = np \text{ and } np \text{ } is \text{ an integer,} \end{cases}$$

where $\lceil \cdot \rceil$ denotes the *ceiling function*. Other terminology relevant to R that appear in this context are: x_k is referred to as the kth *order statistic* in the data, and q_p is referred to as the *sample quantile* corresponding to a given probability p. So, a quantile (another word for a percentile) is a data estimate for a given dataset that corresponds to a given percentile rank.

In R, quantiles are obtained using the function `quantile`, the usage definition of which is

```
quantile(x, probs = seq(0, 1, 0.25), na.rm = FALSE,
    names = TRUE, type = 7, ...)
```

The argument, `type` identifies the algorithm used to compute the desired quantile. Nine algorithms are available, and the algorithm corresponding to the previously stated rule is identified by setting `type = 2`.

For purposes of illustration, the type-2 algorithm is used in the following examples. The first, second and third *quartiles* ($Q_1, Q_2,$ and Q_3, respectively), as defined by the above (type-2) rule are obtained using

```
> quantile(x = alts, probs = c(0.25, 0.50, 0.75),
+     type = 2)
```

Similarly, the first through nineth *deciles* can be obtained using

```
> quantile(x = alts, type = 2,
+     probs = seq(from = 0.1, to = 0.9, by = 0.1))
```

The argument `probs` can be assigned values to determine any set of *quantiles*. For example, using the previously obtained approximate percentile ranks for 365 and 2731, run

```
> quantile(x = alts, probs = c(0.369, 0.901), type = 2)
```

Keep in mind that quantiles computed for a sample using arbitary or approximate percentile ranks will not necessarily be members of the sample.

The default algorithm for the function `quantile` is `type = 7` which computes the quantile corresponding to a given percentile rank, p, through interpolation as follows. For this algorithm the percentile rank of the jth order statistic, x_j, in a given sample is defined by

$$p_j = \frac{j-1}{n-1}.$$

Then, for a given percentile rank p, the order statistics x_j and x_{j+1} are identified for which

$$p_j \leq p < p_{j+1}.$$

It turns out that under this condition $j = \lfloor (n-1)p + 1 \rfloor$, where $\lfloor \cdot \rfloor$ denotes the *floor function*, and the quantile corresponding to p associated with the given sample is computed using

$$q_p = x_j + \left\lfloor \frac{p - p_j}{p_{j+1} - p_j} \right\rfloor (x_{j+1} - x_j).$$

In deciding which `type` to use, it is helpful to note that type-2 is for samples associated with discontinuous variables, and type-7 is for samples associated with continuous variables. For illustration purposes the first, second and third type-7 quartiles for `alts` are computed using

```
> quantile(x = alts, probs = c(0.25, 0.50, 0.75),
+     type = 7)
```

It will be observed that the first and third quartiles computed here differ from the previously computed type-2 quartiles.

Unless otherwise stated, or unless the underlying random variable of a sample is discrete, from here on all quantiles are computed using the default type-7 algorithm.

Standardized Data and z-Scores

For any sample of numeric data, $x_1, x_2, \ldots, x_n$, the transformation

$$z_i = \frac{x_i - \bar{x}}{s}$$

yields the *standardized values* (centered about the sample mean and scaled by the sample standard deviation), or *z-scores* of the data. So, for example, get the z-scores for the entries of `alts` using the standardizing transformation

```
> zScore <- scale(x = alts, center = TRUE, scale = TRUE)
```

Then, the z-score of 300 is

```
> zScore[alts == 300]
```

Because of the nature of the transformation, one unit in the transformed data represents one standard deviation to the left (negative) or to the right (positive) of the sample mean. Thus, in `alts`, 300 is approximately 0.7 standard deviations to the left of the sample mean.[8]

A back-transformations of a given z-score, z, is accomplished using

$$x = \bar{x} + s\,z,$$

where $\bar{x}$ and s are obtained from the sample in question. Keep in mind that, depending on rounding, the results of such back-transformations will not necessarily lie in the sample.

5.3.4   Measures of Shape

The plots contained in Figure 5.1 illustrate five *shape* characteristics in data, and of the probability distributions of the underlying random variables, that are of interest in this section.

The central (reference) shape is that of normally distributed data such as, for example, displayed by the density histogram in Figure 5.1(c), and the normal probability density curves in all five plots. The other four plots illustrate two ways in which data and the distribution of the underlying random variable can deviate from this reference shape.

The top two plots in Figure 5.1 illustrate deviations from symmetry about the mean: Data are either skewed to the left, Figure 5.1(a); or to the right, Figure 5.1(b). The two bottom plots illustrate deviations from normal *peakedness* in data that are symmetric about the mean: The distribution of the data is either flatter about the mean than that of the associated normal distribution, Figure 5.1(d); or more peaked, Figure 5.1(e).

The measures of interest here are *skewness* and *kurtosis*, used to measure deviations in the data from symmetry, and peakedness in relation to the normal distribution (see, for example, [38]). Notation used to denote the various estimates of these measures, and the actual estimates used, vary quite a bit across the literature, see for example [73].

For the following discussion, first establish the terminology, notation and quantitative background used. Let X be a random variable with mean μ.

---

[8] As called here, the `scale` function performs the evaluation

```
(alts - mean(alts))/sd(alts)
```

This function has further capabilities that can be used on matrices. Another function, `sweep`, can be taken advantage of for centering and scaling entries of arrays; see Section 6.2.6.

Then, the *expected value* of X, denoted by $\mathrm{E}(X)$, is the theoretical mean μ of X. Now, for $k \geq 2$, denote the kth *central moment* of X by μ_k, where

$$\mu_k = \mathrm{E}\left[(X - \mu)^k\right].$$

Moments of a random variable are used in computing the skewness and kurtosis of the corresponding probability distribution.

Functions used to compute *sample skewness* and *sample kurtosis* are contained in package `moments` (see also package `timeDate`). Code to perform the same from scratch is given in the script file. Assuming package `moments` is already installed on your computer, load package `moments` in the workspace.

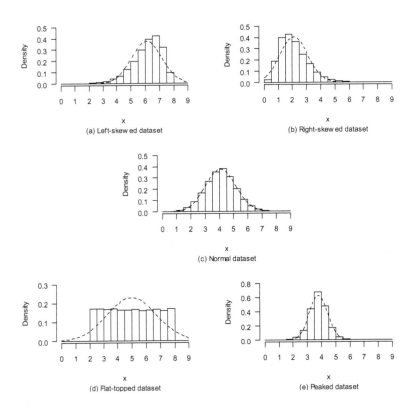

FIGURE 5.1

Density histograms of each of the datasets contained in the list `shapeData`. Superimposed on each of these plots, for reference purposes, is the density curve of the normal distribution having the same mean and standard deviation as the sample in question.

Skewness

The *skewness* of X, defined by

$$\alpha_3 = \mu_3/\sigma^3,$$

measures deviation from symmetry of the corresponding probability distribution (see for example [22, p. 80]).

The *sample skewness* formula used in package `moments`, for a sample of size n, is

$$a_3 = \left[\frac{\sum_{i=1}^{n}(x_i - \bar{x})^3}{n}\right] \bigg/ \left[\frac{\sum_{i=1}^{n}(x_i - \bar{x})^2}{n}\right]^{3/2},$$

where $\bar{x}$ represents the sample mean. The estimate a_3 corresponds to the measure g_1 described in [73] and the measure $\sqrt{b_1}$ described in [38].

Deviations of this statistic from zero indicate deviations from symmetry in the data, and provide information about the probability distribution of the underlying random variable in the following manner: A positive sample skewness indicates the sample is *skewed to the right*, and suggests the distribution of the underlying random variable may have a longer right tail than left tail; a negative sample skewness indicates the sample is *skewed to the left*, and suggests the distribution of the underlying random variable may have a longer left tail than right tail; and a sample skewness of zero indicates the sample is symmetric about the mean, suggesting the underlying random variable may have a distribution that is symmetric about the mean.

Now, to compute (and gather) the sample skewness for each of samples `left`, `normal` and `right` from `shapeData`, see Figures 5.1(a)-(c), run[9]

```
> with(data = shapeData,
+      expr = c(left = skewness(x = left),
+               normal = skewness(x = normal),
+               right = skewness(x = right)))
```

If the data passed into the function `skewness` have missing entries, include the argument assignment `na.rm = TRUE` in the function call.

Based on these measures, it is seen that `normal` contains a sample that is approximately symmetrically distributed about the mean. As for the other two, `left` contains a left-skewed sample and `right` contains a right-skewed sample. Drawing formal conclusions about asymmetry with respect to the underlying random variables requires the performance of formal tests to determine whether a deviation from symmetry is statistically significant. See, for example, [34].[10]

---

[9]The same can be produced using the `sapply` function, see the script file and Section 6.2.1 for more on this useful function.

[10]The function `agostino.test`, from package `moments` can be used to perform the com-

The function `skewness` from package `timeDate` computes the sample skewness using s^3 in place of the denominator used in the previous formula (s being the sample standard deviation of the data being examined).

Kurtosis

The *kurtosis* of X is defined by

$$\alpha_4 = \mu_4/\sigma^4.$$

When applied to symmetric distributions, this quantity measures deviation from normality in terms of the peakedness of the distribution near the mean, and tail weight of the corresponding probability density function (see for example [22, p. 80]).

The *sample kurtosis* formula used in package `moments`, for a sample of size n, is

$$a_4 = \left[\frac{\sum_{i=1}^{n}(x_i - \bar{x})^4}{n}\right] \Big/ \left[\frac{\sum_{i=1}^{n}(x_i - \bar{x})^2}{n}\right]^2.$$

This estimate corresponds to the measure g_2 described in [73] and the measure b_2 described in [38].

For this statistic, deviations from 3 indicate deviations from normal peakedness (or flatness). In [111] it is stated that kurtosis depends on two characteristics of the underlying random variable: how peaked (or flat) the corresponding probability distribution is near the mean; and how heavy (or light) the tails are. Thus, if a symmetric sample yields a kurtosis of 3, then the underlying random variable may have a normal (*mesokurtic*) distribution. For a symmetric sample, if the kurtosis is greater than 3 then the underlying random variable may have a *leptokurtic* distribution, having a sharper peak and heavier tails; and if the sample kurtosis is less than 3 the underlying random variable may have a *platykurtic* distribution, having a flatter peak and lighter tails.

To compute the sample kurtosis for each of `flat`, `normal` and `peaked` in `shapeData`, see Figures 5.1(c)–(e), run[11]

---

putations for such a test. However, be aware that there *may* be a tiny error in the code for this function. If this error is present, the alternatives for one-sided tests are incorrectly reported. To fix this error, run

```
fix(agostino.test)
```

then scroll down to the line

```
else if (alter == 1) {
    alt <- "data have positive skewness"
```

and change the 1 to a 2. Once this is done, close the editor window and save the change.

[11] Again, the same can be produced using the `sapply` function, see the script file and Section 6.2.1.

```
> with(data = shapeData,
+     expr = c(flat = kurtosis(flat),
+              normal = kurtosis(normal),
+              peaked = kurtosis(peaked)))
```

Missing entries are handled in the same way as for the function `skewness`.

Drawing formal conclusions about kurtosis (peakedness) with respect to the underlying random variables requires the performance of formal tests to determine whether a deviation from normal peakedness is statistically significant, see [5].[12]

The function `kurtosis` from package `timeDate` computes the sample kurtosis using s^2 in place of the denominator in the above formula. Before continuing, remove package `moments` from the workspace with the help of the `detach` function.

5.4 Five-Number Summaries and Outliers

For numeric vectors, the extremely versatile `summary` function computes and outputs common statistics that includes the *five-number summary*. This summary can be used as a preliminary indicator of a deviation from symmetry, or of the possible presence of outliers in the data. If the mean is substantially larger than the median, outliers might be present in the right tail, or the sample may simply be skewed to the right; and if the mean is substantially smaller than the median, outliers may be present in the left tail, or the sample may simply be skewed to the left.

Two versions of this summary are presented here: The first is seen in most elementary statistics texts, and for purposes of this book this summary is referred to as the *elementary five-number-summary* (see, for example, [12], [77], or [95]); the second is referred to as *Tukey's five-number-summary* (see, for example, [90]), this is what is computed by the built-in function `fivenum`.

First, to get a couple of samples to use for examples in this section, run the following code.

---

[12]The function `anscombe.test`, from package `moments` can be used to perform the computations for such a test. Be aware that the code for this function *may* also contain a tiny error, similar to that in the function `agostino.test`. If the error is present, the alternatives for one-sided tests are incorrectly reported. To fix this error, run

```
fix(anscombe.test)
```

then scroll down to the line

```
else if (alter == 1) {
    alt <- "kurtosis is greater than 3"
```

and change the 1 to a 2. Once this is done, close the editor window and save the change.

```
> females <- with(data = deerStudy,
+      expr = inland[(month == "7") & (gender=="Female")])
> males <- with(data = deerStudy,
+      expr = inland[(month == "7") & (gender=="Male")])
```

These subsets of the deerStudy data contain the distances from the shore at which male and female deer were observed during the month of July.

5.4.1 Elementary Five-Number Summary

As described in elementary statistics textbooks this summary comprises the minimum value, first quartile, median, third quartile, and maximum value of a numeric univariate dataset. This summary is produced by the function summary (including the mean in the fourth position) when applied to numeric vectors. Keep in mind that the quartiles computed by the summary function are of type-7. If type-2 quartiles are desired, use the function quantile as described in Section 5.3.3 to obtain these. Since the underlying variable for inland is a continuous variable, the following illustrations use type-7 quartiles only.

Observations in a sample are identified as *outliers* if they lie $1.5 \times IQR$ or more units below the first quartile, or above the third quartile. These cutoff values are called the *lower* and *upper fences*. Observations are identified as *extreme outliers* if they lie beyond $3 \times IQR$ units below the first quartile, or above the third quartile, and outliers that are not flagged as being extreme are referred to as *moderate outliers*. Though fairly simple, the outlier-detection method described here is considered robust. See, for example, [3] and [8] for deeper discussions on the topic of outliers in data.

Example: Data with Moderate Outliers

For the vector females the elementary five-number-summary can be obtained by running

```
> (femaleNumbers <- quantile(x = females,
+      probs = c(0, 0.25, 0.50, 0.75, 1)))
```

Using the criteria discussed in the introductory comments of this section, observations flagged as outliers can be identified and listed using code of the following form. First, obtain the interquartile range and then calculate the fences,

```
> iqr <- IQR(females)
> (f <- femaleNumbers[c(2, 4)] + c(-1.5, 1.5)*iqr)
```

Notice that the entry names are inherited from the vector femaleNumbers. Next, to list the outlying values, run

```
> females[(females < f[[1]]) | (females > f[[2]])]
```

It is found that the data values 1.3 and 1.5 represent moderate outliers in the sample; it can be shown that the sample does not contain any extreme outliers.

Example: Data with Extreme Outliers

Now repeat the above computations for `males`, focusing on finding extreme outliers. The five-number summary is

```
> (maleNumbers <- quantile(x = males,
+       probs = c(0, 0.25, 0.50, 0.75, 1)))
```

and the extreme outliers can be found using

```
> iqr <- IQR(males)
> f <- maleNumbers[c(2,4)] + c(-3,3)*iqr
> males[(males < f[[1]]) | (males > f[[2]])]
```

So the sample `males` contains five extreme outliers in the left tail of the data, 1.8, 1.9, 2.0, 2.1, and 2.2; there are two moderate outliers in the right tail, 3.2 and 3.3.

5.4.2   Tukey's Five-Number Summary

The statistics for this summary are computed by the function `fivenum`, and typically differ from the previous (elementary) five-number summary if the sample size, n, is an even number. The samples `females` and `males` have odd numbered lengths so, as should be expected,

```
> fivenum(females); fivenum(males)
```

will produce the same numbers as found in the previous two examples. Outliers in the data can be flagged using the approach described for the elementary five-number summary.

For this (Tukey's) summary, the first and third quartiles from the elementary five-number-summary are replaced by what are referred to as the lower and upper hinges, respectively. The *lower hinge* is the $(\lfloor (n+1)/2 \rfloor + 1)/2$ order statistic, where $\lfloor \cdot \rfloor$ denotes the *floor function*; and the *upper hinge* is the $(\lceil (n+1)/2 \rceil + n)/2$ order statistic, where $\lceil \cdot \rceil$ denotes the *ceiling function*. If the computed index for an order statistic is not an integer, the mean of the adjacent order statistics is computed. These hinges equal the first and third quartiles for odd n; and for even n they are close in value, or even equal to the two quartiles. The added benefit of the use of these hinges is that they equal observations in the data more often than quartiles do.

5.4.3 The `boxplot.stats` Function

The function `boxplot.stats` uses the function `fivenum` to compute and list outlying observations using the $1.5 \times IQR$ criteria (by default), with IQR being calculated using Tukey's upper and lower hinges. For example, to list all possible outliers in `females`, run

```
> boxplot.stats(x = females, coef = 1.5,
+     do.conf = FALSE)
```

and to list only extreme outliers in `males`, run

```
> boxplot.stats(x = males, coef = 3,
+     do.conf = FALSE)
```

The lowest and highest values listed under `$stats` are the most extreme data values that lie within (or on) the lower and upper fences. These are the values to which the boxplot *whiskers* extend (see Section 8.2), the lengths of which are determined by the value assigned to the argument `coef`. Observations outside of these fences are identified as outliers under `$out`.

The assignment `do.conf = FALSE` suppresses the computation of a confidence interval of the median using the sample in question; see Section 8.2.

6

More on Computing with R

The time has come to revisit computations with vectors in a little more detail, and introduce some additional functions, ideas and points to keep in mind when working with multivariate data. It is also useful, at this point, to introduce conditional statements, looping structures, and the fundamentals of building functions, see code in `Chapter6Script.R`.

6.1   Computing with Numeric Vectors

Since the names of the vectors can be viewed as variables, they can be placed in a formula for purposes of computations. There are, however, some points to be aware of when using vectors in computations.

Using Single Vectors in a Formula

Suppose the dependent variable y is defined by

$$y = 3 - 2x + x^2.$$

Then, a set of ordered pairs (x, y) can be obtained and displayed using, for example,

```
> x <- 0:5; y <- 3 - 2*x + x^2; rbind(x, y)
```

Each entry of the vector y is obtained by term-wise computations on the entries of x. That is, for each $i = 1, 2, \ldots, 5$,

$$y_i = 3 - 2x_i + x_i^2.$$

So, as long as each entry of y is a well-defined real number corresponding to each entry of the vector x, there are no worries.

Working with Vectors of Equal Lengths

Now suppose two independent variables, say x and y, are combined to define the dependent variable z by

$$z = 3 + 2x - 3y + xy - x^2 + y^2.$$

Generate values for x and y using, for example,

```
> x <- seq(from = 0, to = 5, length.out = 10)
> y <- seq(from = 15, to = 5, length.out = 10)
```

Then,

```
> z <- 3 + 2*x - 3*y + x*y - x^2 + y^2
```

computes the corresponding values for z, and the entries can be viewed in a manner similar to the previous example.

The entries of the resulting vector z are again obtained following the term-wise approach. That is, for $i = 1, 2, \ldots, 10$,

$$z_i = 3 + 2x_i - 3y_i + x_i y_i - x_i^2 + y_i^2.$$

Here too, there are no worries as long as the formula used to compute z is well-defined, and produces a real number for each pair of entries from the vectors x and y.

When Vectors Have Unequal Lengths

This has already occurred in the previous illustrations at a basic level. For example, in the calculation

```
> z <- 3 + 2*x - 3*y + x*y - x^2 + y^2
```

the constant term 3 and the coefficients of x and y on the right side are vectors of length 1. Since the lengths of the vectors x and y are integer multiples of 1, the numbers 3 and 2 are recycled for operations on each entry of x and y.

Of interest is what happens when vectors of unequal lengths (greater than 1) are involved in a computation. Consider three vectors, all of differing lengths, for example,

```
> x <- c(1, 2); y <- c(1, 2, 3, 4); z <- c(1, 2, 3, 4, 5)
```

and first look at the sum of x and y.

```
> x + y
[1] 2, 4, 4, 6
```

Since no warnings are issued, this seems like a valid computation. Here is what happened.

The first two entries of x + y are the sums of the first two entries of x and y, then since the length of y (which is 4) is an integer multiple of the length of x (which is 2), the entries of x are recycled. That is

$$x + y = 1 + 1, \ 2 + 2, \ 1 + 3, \ 2 + 4.$$

It is worth mentioning that such computations (unless intentional) are likely to lead to misleading or possibly incorrect results, and since no warnings are issued it is unlikely potential errors would be expected or noticed. Now look at what happens if the sum of x and z is computed.

```
> x + z
[1] 2 4 4 6 6
Warning message:
In x + z : longer object length is not a multiple
of shorter object length
```

While the computation is performed (the terms of x are recycled as before), a warning message does alert the user to a matter of possible concern.

A Takeaway Message

So, a useful caution when computing with vectors is to make sure all of the vectors involved have equal length. The simple exceptions to this rule are scalar operations such as those described in the beginning of this section. Alternatively, if you know exactly what you are doing, exactly what you want to get, and if you are very comfortable with coding, you may be able to take advantage of this recycling feature.

6.2 Working with Lists, Data Frames, and Arrays

The functions demonstrated here permit the performance of repetitive tasks, such as those that typically require a loop to perform. The data frame used for the first few demonstrations in this section is obtained as follows. First run

```
> july <- subset(x = deerStudy, subset = (month == "7"),
+      select = c(subject, year, gender, island,
+          weather, altitude, weight))
```

to extract data under subject, year, gender, island, weather, altitude, and weight for the month of July from the data frame deerStudy and then rename the rows of the resulting data frame.

6.2.1 The sapply Function

This function can be used to perform operations across the contents of a list containing factors, or vectors having the same class. The usage definition for this function, with arguments of interest, is

```
sapply(X, FUN, ...)
```

The argument X has to be a list of vectors (or factors), or an object that can be coerced into a list of vectors (or factors). The argument FUN is the function that is to be applied to the contents of X, and the argument(s) following FUN, denoted by "...," are those that may be needed by the function assigned to FUN.

Prepare a list to work with, for example, one containing four vectors of altitudes at which deer were observed on each island. The split function finds use here,

```
> islandAlts <- with(data = july,
+      expr = split(x = altitude, f = island))
```

Now for some examples.

When the Output of FUN Is Single-Valued

If the function assigned to the FUN argument is such that it produces single-valued output for any vector (or factor) it is applied to, then the sapply function produces a vector. For example,

```
> sapply(X = islandAlts, FUN = max)
Admiralty  Baranof Chichagof  Douglas
    2814     2834      2831      2378
```

A variety of single-valued functions can be used in the sapply function in this manner to produce meaningful output.

When the Output of FUN Is a Vector

If the function assigned to the FUN argument is such that it produces vector output for any vector (or factor) it is applied to, then the sapply function produces a matrix.

There are occasions when both the rows and the columns of the resulting matrix are named. For example,

```
> sapply(X = islandAlts, FUN = quantile,
+      probs = c(0.25, 0.75))
        Admiralty Baranof Chichagof Douglas
25%      1485.5 1445.75   1641.75  1508.5
75%      2471.0 2521.50   2554.25  2014.0
```

Here the argument assignment probs = c(0.25, 0.75) is an example of the "..." part of the function call.

Depending on the function used, the rows outputted may not be assigned names. For example, run

```
> sapply(X = islandAlts, FUN = range)
```

If row names are desired, one approach might be to take advantage of the data.frame function as follows.

```
> data.frame(sapply(X = islandAlts, FUN = range),
+     row.names = c("lower", "upper"))
```

Again, a variety of vector-valued functions can be used in the sapply function in this manner.

Using sapply on Data Frames

Since data frames are of class list, as long as the columns of interest have the same type, or the function assigned to FUN "knows" what to do with columns of different types, the sapply function can be applied to data frames. For example, mean weights and altitudes from the july data frame can be obtained using

```
> sapply(X = july[ , c("weight", "altitude")],
+     FUN = mean, na.rm = TRUE)
```

Recall that the na.rm = TRUE assignment is needed to address (remove) the NA entries in weight; however, the corresponding (non-NA) entries in altitude are left untouched.

Graphics functions and a variety of hypothesis test functions can also be used in sapply. Examples of such applications and others appear in the script files for later chapters.

6.2.2   The tapply Function

The usage definition of this function with arguments of interest is

```
tapply(X, INDEX, FUN = NULL, ...)
```

and it operates in much the same manner as the sapply function, except that in this case X is a single vector (or factor), and the INDEX argument provides a means of partitioning the data by levels of one or more factors within the tapply function without having to extract a list of separate vectors. The argument assignment na.rm = TRUE can also be used as it is in the sapply function.

Unlike the sapply function, the tapply function does not always simplify output into matrix form. Rather, for cases when the function assigned to FUN produces two or more values for each transect, the output appears as a list. As shown in the script file, it is a simple matter to convert these lists into matrix format. Here are some examples.

Computations by Levels of a Single Factor

If the function assigned to the FUN argument is such that it produces single-valued output for any vector (or factor) it is applied to, then the `tapply` function produces a vector. For example, the first demonstration in the previous section can be duplicated using

```
> with(data = july,
+      expr = tapply(X = altitude, INDEX = island,
+          FUN = max))
```

Similarly, the second and third examples in the previous section can be duplicated, but in list format as opposed to matrix format; see the script file for one way in which the output can be reformatted.

```
> with(data = july,
+      expr = tapply(X = altitude, INDEX = island,
+          FUN = quantile, probs = c(0.25, 0.75)))
```

and

```
> with(data = july,
+      expr = tapply(X = altitude, INDEX = island,
+          FUN = range))
```

As in the case of the `sapply` function, a variety of functions can be used in the `tapply` function in this manner.

Computations by Levels of Two or More Factors

If the function assigned to FUN is single-valued, computations can be performed and outputted by the levels of two or more factors. For example, to get mean altitudes by gender and island, run

```
> with(data = july,
+      expr = tapply(X = altitude,
+          INDEX = list(gender, island), FUN = mean))
```

Note that if `weight` is used in place of altitude in any of the previous applications of the `tapply` function the argument assignment `na.rm = TRUE` needs to be included to ensure means are computed.

To get mean altitudes by gender on each island, *and* for each year, run

```
> with(data = july,
+      expr = tapply(X = altitude,
+          INDEX = list(gender, island, year), FUN = mean))
```

The result is an array containing four tables, one for each year.

If the function assigned to FUN is *not* single-valued, computations by the levels of two or more factors are still performed; however, the results of the computations are not necessarily displayed in a convenient format. Some work has to be done to get at the contents of the resulting object. See the script file for exploratory code.

6.2.3   The by Function

The by function has the following usage definition,

```
by(data, INDICES, FUN, ..., simplify = TRUE)
```

and serves the same purpose as the tapply function. The output for this function can be a list or an array.

For simpler cases, such as applying a single-valued function by levels of one or two factors, the tapply function is probably preferable as the output is in vector or matrix form and the contents of the resulting object are more conveniently accessed.

When FUN is Assigned a Single-Valued Function

For example, while the tapply produces a vector when the function max is applied to altitude by island in the data frame july, the by function produces a list; see the output obtained by running

```
> with(data = july,
+     expr = by(data = altitude,
+           INDICES = island, FUN = max))
```

Deciding between using the by or tapply function may be based on what the output is to be used for.

When FUN Is Assigned a Multi-Valued Function

If the function assigned to FUN is not single-valued, and the computations are to be performed by the levels of two or more factors, then the by function may be preferred over the tapply function. For example,

```
> (altRanges <- with(data = july,
+     expr = by(data = altitude,
+           INDICES = list(gender, island), FUN = range)))
```

produces and displays a list of altitude ranges organized by the levels of gender and island.

It is helpful to know how to access the contents of the resulting list. The `dim` function, when applied to `altRanges`, shows there are two dimensions, one of length 2 and the other of length 4. Then, using the function `dimnames` it is found that the first dimension contains the genders, and the second the island names. So, for example, to extract the range of altitudes for females on Chichagof island, one can run

> `altRanges[[2,3]]`

Alternatively, dimension names can be used to extract the contents.[1]

> `altRanges[["Female", "Chichagof"]]`

Another approach can be used to access the contents of `altRanges`. Running `length(altRanges)` suggests there are 8 objects in `altRanges`. The previously determined dimensions (2×4) suggests that the first two contain the ranges for males, then females on Admiralty; the next two for males, then females on Baranof and so on. So, to extract the range of altitudes for females on Chichagoff island, the 6th entry is needed,

> `altRanges[[6]]`

Again, a variety of functions can be used in the `by` function, and as with the `sapply` and `tapply` functions, the argument assignment `na.rm = TRUE` may be needed if `NA` entries are present in the data.

6.2.4 The `aggregate` Function

This function can be used to get equivalent, but differently formatted output as from the `tapply` and `by` functions. It also enables operations on columns of a data frame by levels of multiple factors, but then outputs the results as a data frame. The usage definition for the `aggregate` function is

```
aggregate(formula, data, FUN, ...,
        subset, na.action = na.omit)
```

The argument `formula` is what determines the manner in which the computations are performed by the function assigned to `FUN` on the `data` supplied. By default, all rows containing `NA` entries are removed from the `data`. Here are some illustrations.

Computations on One Variable by Levels of One or More Factors

First, consider obtaining the mean `weight` (with `NA` values removed, by default) of all deer observed in July by `gender` and `island`.

---

[1] The use of `[[ ]]` is made so that only the contents are extracted.

```
> (meanWt <- aggregate(formula = weight ~ gender + island,
+       data = july, FUN = mean))
```

The object created, `meanWt`, is a data frame and its contents can be accessed using methods described in Section 3.4.

The `subset` argument can be used to get summaries for subsets of the data. For example, to get the mean `weight` of deer observed in July by `gender` and `island` below an altitude of 2000 feet, run

```
> aggregate(formula = weight ~ gender + island,
+       data = july, FUN = mean, subset = (altitude < 2000))
```

While the output has fewer rows, the columns are the same as those of the previous example.

Computations on Two Variables by Levels of One or More Factors

To compute means of `altitude` and `weight` by `gender`, use

```
> aggregate(formula = cbind(altitude, weight) ~ gender,
+       data = july, FUN = mean)
```

Going further, compute means of `altitude` and `inland` by `gender` and `island` using

```
> aggregate(formula = cbind(altitude, weight) ~ gender +
+       island, data = july, FUN = mean)
```

Subsets of the data can be analyzed as described for the one variable case.

Another Way to Get Multi-Dimensional Contingency Tables

The `aggregate` function can also be used to obtain frequency distributions by the levels of three or more factors in a more compact form (depending on the number of factors involved) than the `table` function described in Sections 5.1 and 5.2. For example, frequencies of deer observed by `gender`, `island` and `weather` can be obtained using

```
> aggregate(formula = subject ~ gender + island + weather,
+       data = july, FUN = length)
```

The resulting outputted object is a data frame, compare the output with that of

```
> with(data = july, expr = table(gender, island, weather))
```

where the resulting object is a $2 \times 4 \times 3$ array.

As with the previously illustrated `sapply`, `tapply` and `by` functions, a variety of functions can be used in the `aggregate` function.

6.2.5   The `apply` Function

This function takes in an *array* of data,[2] and applies a given function to a chosen *margin* (such as row, column, or other), or a collection of margins. The object `mileTimes` is used for the following illustrations; here is a description of the contents of this array.

Three friends, Sam, Sal, and Sol, decided to start training for a road race and, every day, over a three week period they recorded their times for a one-mile run. The data are placed in a 3-dimensional array; see the script file for how this is done.

Of interest at this point is the matter of navigating such an array. The contents of an array are accessed by indices, or dimension level names, in much the same manner as those of a matrix or a data frame. The dimensions of the array in question are $3 \times 7 \times 3$, where the first dimension represents the weeks, the second represents the days, and the third the individuals.

So, for example, Sal's times for each day of week 2 can be accessed using either of

```
> mileTimes[2, , 2]
> mileTimes["Week2", , "Sal"]
```

Similarly, Sam's and Sol's times for the Sundays of all three weeks can be extracted using either of

```
> mileTimes[ , 7, c(1, 3)]
> mileTimes[ , "Sun", c("Sam", "Sol")]
```

The usage definition of the `apply` function is as follows,

```
apply(X, MARGIN, FUN, ...)
```

where the arguments `X` and `FUN` have the same meaning as for the `sapply` and `tapply` functions, and as before "..." represents arguments used by the function assigned to `FUN`. The `MARGIN` argument is used to identify the dimension(s) of interest in the array. For example, using the `mileTimes` array, `MARGIN = 1` identifies the weeks as being of interest, `MARGIN = 2` the days of the week, and `MARGIN = 3` the individuals.

Computations can be performed over one or more margins, and the output can be a vector, a matrix, or an array depending on the dimensions of the array assigned to `X` and/or what the function assigned to `FUN` outputs.

Computations Over a Single Margin

To find the mean times for Sam, Sol, and Sal over the whole three week period, run

---

[2]Since vectors are 1-dimensional arrays and matrices are 2-dimensional arrays, these can also be operated on using the `apply` function.

```
> apply(X = mileTimes, MARGIN = 3, FUN = mean)
```

Similarly, only Sam's weekly averages can be obtained using

```
> apply(X = mileTimes[ , , "Sam"], MARGIN = 1, FUN = mean)
```

Note that `mileTimes[ , , "Sam"]` is a 2-dimensional array (a matrix), and contains only Sam's running times.

Computations Over Two Margins

To look at weekly mean times for Sam, Sal, and Sol for each of the three weeks, run

```
> apply(X = mileTimes, MARGIN = c(1, 3), FUN = mean)
```

Similarly, daily averages of all three runners' times for each of the three weeks can be obtained using

```
> apply(X = mileTimes, MARGIN = c(1, 2), FUN = mean)
```

For a 3-dimensional array the most number of margins that is typically assigned to `MARGIN` is two. This general rule is relaxed if the function assigned to `FUN` tranforms the data from one scale to another, returning an array of the same dimensions. This generalizes to any array of two or more dimensions. Here are some additional points to keep in mind.

The argument `MARGIN` is assigned the integers that identify the coordinates of interest (such as the "x-coordinate," the "y-coordinate," and so on). Thus, for an n-dimensional array, `MARGIN` can be assigned any single value from 1 through n, or a vector containing any combination of these numbers.

Note that when working with matrices, a more efficient method of computing row and column means is to use the functions `rowMeans` and `colMeans`. Similarly, there are the functions `rowSums` and `colSums`.

6.2.6 The `sweep` Function

The usage definition, with arguments of interest, is

```
sweep(x, MARGIN, STATS, FUN = "-", ...)
```

where `x` is the array on which the sweeping is to be carried out by a two argument function assigned to `FUN` (the default being subtraction), using the summary statistic assigned to `STATS` in the margin(s) assigned to `MARGIN`. Any optional arguments needed by the function `FUN` are passed in as optional arguments through "...".

Understanding how this function works and how to use it requires some concentration. First, the function assigned to `FUN` takes in the arrays assigned to `x` and `STATS` as the arguments on which it is to act. So, particular attention has to be paid to the dimensions of STATS in relation to x, and the margins of x that are to be swept.

Here are some examples using the array mileTimes. Begin by letting x_{ijk} represent the observed time of individual k (out of Sam, Sal and Sol) on day j (out of Monday through Sunday) in week i (out of 1 through 3).

Sweeping Along a Single Margin

Let $\bar{x}_{..k}$ denote individual k's overall mean time for the whole three-week period and suppose the wish is to compute

$$x_{ijk} - \bar{x}_{..k}.$$

The sample means, $\bar{x}_{..k}$, for the three individuals are easily computed and examined, then the desired end-result is obtained using

```
> (overallMean <- round(apply(X = mileTimes, MARGIN = 3,
+        FUN = mean), digits = 3))
> sweep(x = mileTimes, MARGIN = 3, STATS = overallMean)
```

The results can be checked quite easily; see the script file.

Now focus on the dimensions of mileTimes and overallMean, and the margin of mileTimes along which the sweep is conducted. The dimensions of mileTimes are $3 \times 7 \times 3$, and the margins represent week, day of week and individual name. Since the margin of mileTimes along which the sweep is to be conducted is 3 (times separated by the three individuals), and the contents of overallMean should correspond to each of the three individuals, overallMean should be a vector of length 3. So, for example, Sam's overall mean is subtracted from each of Sam's times.

Sweeping Along Two Margins

Now let $\bar{x}_{i\cdot k}$ denote individual k's overall mean time for week i and suppose the wish is to compute

$$x_{ijk} - \bar{x}_{i\cdot k}.$$

The sample means, $\bar{x}_{i\cdot k}$, for the three individuals for each week is first computed, and then the desired sweep is performed using

```
> (weeklyMean <- round(apply(X = mileTimes,
+    MARGIN = c(1,3), FUN = mean), digits = 3))
> sweep(x = mileTimes, MARGIN = c(1, 3),
+    STATS = weeklyMean)
```

Again, the results can be verified.

Once again, the dimensions of `mileTimes` are $3 \times 7 \times 3$, and the margins represent week, day of week and individual name. Since the margins of `mileTimes` along which the sweep is to be conducted are 1 (week) and 3 (individual name), the contents of `weeklyMean` should correspond to not only each of the three individuals but also each of the three weeks. So, `weeklyMean` should be a 3×3 matrix and, for example, for each week Sam's weekly mean is subtracted from each of Sam's times within the week in question.

Sweeping by a Function Other than "-"

The key restriction on the function assigned to `FUN` is that it takes in the arrays assigned to `x` and `STATS`, and operates on these two arrays. This provides for quite a bit of flexibility. Here is a fairly simple example.

Let $s_{i \cdot k}$ denote the standard deviation of individual k's times for week i and suppose the wish is to compute

$$x_{ijk}/s_{i \cdot k}.$$

That is, scale the times for each day by the individual's corresponding weekly standard deviation. The sample standard deviations, $s_{i \cdot k}$, for the three individuals for each week is computed using the `apply` function,

```
> weeklySd <- round(apply(X = mileTimes,
      MARGIN = c(1,3), FUN = sd), digits = 3)
```

and the scaling is accomplished by running

```
> sweep(x = mileTimes, MARGIN = c(1, 3),
      STATS = weeklySd, FUN = "/")
```

Again, the results can be verified. See the script file for an exploratory example in which the function applied is user-defined.

A Sweep Within a Sweep

Consider using the `sweep` function to center and scale each individual's times by the corresponding weekly averages and standard deviations; that is, compute the standard scores

$$(x_{ijk} - \bar{x}_{i \cdot k})/s_{i \cdot k}.$$

The two matrices `weeklyMean` and `weeklySd` can be used in a double application of the `sweep` function as follows. First center the data about each individual's weekly means,

```
> centered <- sweep(x = mileTimes,
+      MARGIN = c(1, 3), STATS = weeklyMean)
```

then scale the results by each individual's weekly standard deviations.

```
> sweep(x = centered, MARGIN = c(1, 3),
        STATS = weeklySd, FUN = "/")
```

Note that a more efficient way to perform this task would be to use the scale function within the apply function; see the script file.

6.3 For-Loops

For most cases, applications of the various "apply" functions (as well as the by and aggregate) are the most efficient way to perform tasks that lend themselves to looping constructs. However, when none of these functions can be easily applied to a set of repeatable tasks, the for-loop provides an effective alternative.

A key ingredient in using for-loops is when a particular predetermined set of tasks is to be performed on a known collection of objects of the same type and class. The syntax for a for-loop can be summarized as follows:

```
for (index in indexVector)
{   # Start loop body
    # For each index, perform given tasks

    :

}   # End loop body
```

The contents of the index vector can be integers or character strings (such as names of objects), and the tasks coded into the body of the loop are performed on each object or its contents as identified by the index.

A Single Loop

The task performed in the second example of Section 6.2.1, using the sapply function, can be duplicated using the following code. First prepare an index vector and a matrix to receive the results.

```
> islands <- names(islandAlts)
> altRanges <- matrix(nrow = 2, ncol = 4,
+       dimnames = list(c("lower", "upper"), islands))
```

Next, run

```
> for (i in islands)
+ {   # Start loop body
+     altRanges[ , i] <- range(islandAlts[i])
+ }   # End loop body
```

The results can then be checked by running `altRanges`.

In this example the vector `1:4` can be used as the index vector in place of the character string vector `islands`.

Nested Loops

The computations performed for the fourth example in Section 6.2.2 can be duplicated using *nested for-loops* as follows. First, attach the data frame `july` for convenience, then prepare the index vectors, and define a matrix to receive the results of the tasks performed.

```
> genders <- levels(gender); islands <- levels(island)
> altMeans <- matrix(nrow = 2, ncol = 4,
+     dimnames = list(genders, islands))
```

Then, the nested loops to extract the mean altitudes by gender and island are as follows.

```
> for (i in genders)
+ {   # Start outer loop body
+     for (j in islands)
+     {   # Start inner loop body
+         altMeans[i, j] <- mean(
+             altitude[(gender == i) & (island == j)])
+     }   # End inner loop body
+ }   # End outer loop body
```

To wrap things up, detach the data frame `july` and then take a look at the results.

Here, the vector `1:2` can be used for the index `i` in place of `genders`, and the vector `1:4` for the index `j` in place of `islands`.

6.4 Conditional Statements and the `switch` Function

Two constructs that can be useful in coding are the *if* and *if-else* constructs. If used correctly, these enable the inclusion of considerable control on the flow and direction of tasks to be performed.

Though the following examples focus mostly on the use of conditional statements within user-defined functions, keep in mind that they can be used whether or not the code is for a function.

6.4.1   The if-then Statement

In a nutshell, this construct instructs R to perform a specified task (or collection of tasks) if a specific condition is satisfied (or logical statement is true). The general layout for this construct is

```
if (condition)
{    # Begin body of 'if' statement
     # If condition is TRUE, then perform task(s)
     ⋮
}    # End body of 'if' statement
```

While this format covers all applications, there are situations when braces, {...}, need not be included.

Examples Where Braces Are Not Needed

In cases where code for the task is short, the whole statement can be placed on a single line and the opening and closing braces can be left out. Consider, for example, the line

```
if (missing(x) | missing(g)) stop("Missing x or g")
```

from the illustration of constructing a user-defined function in Section 6.5. The objects x and g are required arguments for the function leveneType.test. Here is what this line of code does. If either one of these arguments is missing (not assigned an object) in a call of the function leveneType.test, print the message "Missing x or g" to the screen and stop (exit) the function.[3] Since the stop function call lies on the same line as the if, it does not need to be enclosed by braces.

At times code for an if statement cannot be elegantly placed on a single line, but involves a *single* function call or a *single* (but complex) computation. In such cases it is tidier to begin the code for the task on a new line, here too the opening and closing braces can be left out. For example, again in the code for the function leveneType.test from Section 6.5, the two lines

```
if (!(center %in% c("mean", "median", "trimmed.mean")))
    stop("Incorrect 'center' assignment")
```

---

[3]See also the **warning** function which is useful for issuing warnings to users of a function.

instruct R to issue an error message and exit the function if the method assigned to the optional argument `center` is not one of the three listed. Here too there is no ambiguity with respect to what task is to be performed if the condition is satisfied, so the `stop` function call need not be enclosed by braces.

Sometimes a function call, or code for a single complex computation within an if statement runs over two or more lines. Here again is a situation where the opening and closing braces can be left out.

Suppose part of the code in a function includes the construction of a density histogram *if* the sample in question is large enough for a histogram to be meaningful. The term "large" is somewhat nebulous; for our purposes, let us say samples of size less than 20 do not provide for meaningful histograms. Suppose that, within the function, the sample in question is accessed through an argument named `observations`. Then,

```
if (length(observations) >= 20)

    hist(x = observations, freq = FALSE,

        include.lowest = TRUE, right = FALSE,

        main = paste("Density Histogram of",

            substitute(observations)))
```

ensures that a histogram is plotted only if the sample size is at least 20.

To be on the safe side, include opening and closing braces for any case that does not fit clearly in one of these three scenarios.

An Example Where Braces Are Needed

Section 10.5 provides an example of a user-defined graphics function, `errorBar.plot`; see also Section 9.3. Like the previously mentioned `leveneType.test` function, this function is provided a numeric vector, `y`, and a factor `g`. Then, based on which of four character strings is assigned to an argument called `type`, the function plots one of four types of error bars.

One of the types involves using the interquartile range to determine bounds for the error bars about sample medians. The statistics needed for this type of error bar plot differ from those of the other three, hence the inclusion of a statement of the form

```
if (type == "iqr")
{   # Begin body of 'if' statement
    centers <- tapply(X = y, INDEX = g, FUN = median)
    iqrs <- tapply(X = y, INDEX = g, FUN = IQR)
    bounds <- cbind(centers - iqrs/2, centers + iqrs/2)
}   # Close body of 'if' statement
```

So, if `type == "iqr"` is true, this part of the code obtains the medians, interquartile ranges, and then the error bar bounds for the entries of y by levels of g. If not, this part of the code is by-passed.

As a general rule, when more than one distinct task (or function call) is to be performed when the condition is satisfied, enclose the collection of tasks and/or function calls within braces.

6.4.2   The if-then-else Statement

While this construct also instructs R to perform a specified task (or collection of tasks) if a specific condition is satisfied, it provides for an alternative when the if-condition is not satisfied. The general layout for this construct is

```
if (condition)
{   # Begin body of 'if' part
    # If condition is TRUE, then do this

    ⋮

} else { # End 'if', begin 'else' part
    # If condition is not TRUE, then do this

    ⋮

}   # End body of 'else' part
```

This format covers all applications. It is important to pay close attention to the placement of the closing brace, }, for the `if` part; this should always appear before, and on the same line as the `else`. The opening brace, {, for the `else` part can be placed on the same lines as the `else`, or on the next line.

As with the if statement there are situations when the braces need not be included, but be cautious. Here are some examples.

Examples Where Braces Are Not Needed

The same general rule for parentheses as for the previous if-statement apply, either only for the `if`, or only for the `else` part, or for both the `if` and the `else` parts. Here are three generic examples.

Suppose a particular function has the numeric vector x as its required argument, and another numeric vector y as an optional argument. Now suppose this function contains the code

```
if (missing(y)) hist(x) else plot(x, y)
```

Then, if the argument y is missing in the function call a frequency histogram of the data in x is plotted, otherwise a scatterplot of y against x is plotted.

Using the same idea, suppose the `if` and `else` parts cannot be nicely placed on a single line as shown above; more arguments are to be passed into the functions `hist` and `plot`. Then the code might be altered to look something like

```
if (missing(y))
      hist(x, freq = FALSE, right = FALSE)
else
      plot(x, y, xlim = c(floor(min(x)), ceiling(max(x))))
```

Again, depending on whether y is missing (or not) in the function call there is no ambiguity about what task is to be performed.

As with the `if` statement, braces would still not be needed if either or both of the function calls for `hist` or `plot` extend over more than a single line

An Example Where Braces Are Needed

Going back to the last example for the if statement from the `errorBar.plot` function, the `if (type == "iqr")` is actually tied to an `else`. Here, if `type == "iqr"` is not true, then it must be one of the other three types of error-bar plots (identified by the types `"sd"`, `"se"` or `"ci"`). These are the types that the `else` component of this particular if-else construct addresses through an application of the `switch` function; see code in the script for this example.

So, for example, if `type == "sd"` then the function skips the `if` part of the statement and goes on to the `else` part where it computes bounds based on the sample standard deviations.

6.4.3   The `switch` Function

The `switch` function has the following usage definition

```
switch(EXPR, ...)
```

where `EXPR` is assigned either a character string or an integer that is then used to identify the method to be employed for an evaluation. The evaluation may produce a number or character string. The argument "`...`" is assigned a list of alternatives that `EXPR` can belong to along with the corresponding computational rule.

Here is a line by line explanation of what happens in the computation of error bar bounds through the `switch` function call in the previous example. The argument `EXPR` is assigned a plot `type` which can be one of `"sd"`, `"se"` or `"ci"`. Then:

- if `type == "sd"`, compute the bounds using the formula

$$\text{centers} \pm s;$$

- if type == "se", compute the bounds using the formula

$$\text{centers} \pm \text{s/sqrt(ns)};$$

- if type == "ci", compute the bounds using the formula

$$\text{centers} \pm \text{qt}(0.975, \text{ns} - 1) * \text{s/sqrt(ns)}.$$

It is worth remembering that a nested if-else statement can be used to perform a similar (or the same) tasks; see this section's the script for sample code.

6.5   Preparing Your Own Functions

Some examples of user-defined functions have already appeared in the script files for earlier chapters (Sections 2.6 and 4.2), so this is a suitable place to provide a few more details on what is involved.

Very briefly, an R function contains code for a set of rules/tasks (computation and/or other) that are applied to one or more input items to produce desired output.

General Structure

The following general structure suggests one way in which the various components of a function might be organized.

```
name.task <- function(Required args, default args, ...)
{  # Begin function body
   # Argument checks
   ⋮
   # Initializations
   ⋮
   # Computations
   ⋮
   # Output
}  # End function body
```

Here is a discussion of each component outlined above.

Function name: As pointed out in [6], there does not appear to be a consistent naming convention for R functions. The use of a period in function names appears to be popular, and the convention mentioned in Section 1.10 is one possible suggestion.

Required arguments: These are those objects that the function needs from outside in order to successfully perform its intended tasks. Typical among these are datasets, or other very specific instructions.

Default and optional arguments: These (including "..." if so desired) are those objects that are assigned default values if they are not passed into the function from outside (omitted in a function call). So these objects need not be passed into the function if the default values are acceptable.

Argument checks: While optional, these can be important *if someone else is going to be using your code*. These make sure there are no missing arguments (among the required arguments), and they can also be used to make sure the objects assigned to all the arguments are of the correct structure.

Initializations: These can sometimes be useful to reduce the bulk of later code, and they are particularly important if logical variables are needed for later decisions. These can also be important if a for-loop is to be used to perform some task on one or more variables (see earlier section on for-loops for some simple scenarios).

Computations: Code for computations (or other tasks) are typically designed in a manner much the same as is done outside of a function. The difference here is that the environment in which the computations (which include tasks like graphics) are performed is controlled by what is passed into the function.

Output: This depends a lot on the purpose of the function. Very broadly speaking, output may be to the console, or to a text file, or to a graphics file, or to the graphics device window. In some cases, the results of a function call may feed into another function call.

An Example

The function `leveneType.test` from Section 13.4, designed to perform tests of the equality of variances for two or more populations provides illustrations of each of the above components. See details in the complete documented code for this function contained in the script for this section.

Once a function has been prepared, it is important to test it to make sure there are no glitches in it. Run the code in the script file for the function `LeveneType.test`, then work through the lines of testing code to see that the various argument checks work as they are intended.

Some Suggestions

For the present there are a couple of points to keep in mind when constructing a function. First and foremost, all *variable* information should be passed into the function as arguments. Except for built-in R functions and other built-in R objects, do not use any user-defined objects external to the function in question within the body of the function. If such objects are needed, either pass them into the function as arguments or define/build them into the body of the function.

Second, test code to be used in the body of a function *before* enclosing it in the function. As an illustration of this, see the script file for an example of test-running a computation.

If no errors are found, the code can be included in the function body and tested again by calling the function in a variety of ways. It is much easier to trace code and track down errors outside of the function. It is also useful to use simple (small) datasets with known output/results when testing code for computational correctness.

7

Basic Charts for Categorical Data

This chapter illustrates the use of three high-level graphics functions from package `graphics` for displaying categorical data, and provides an introduction to some of the graphics image formatting and enhancement tools available in R. Also included are several methods for exporting graphics images from R into a variety of file formats. Open the script file `Chapter7Script.R` to follow along.

7.1 Preliminary Comments

Two common approaches to displaying categorical data graphically are presented in this chapter. In the first approach, observed frequencies are represented linearly (for example, by the length of a *bar*, or the location of a *dot*); and in the second, observed frequencies are represented by area (for example, by the area of a sector of a circle or *pie*). The majority of charts used for displaying categorical data may be considered incarnations of one or the other of these two approaches. *Bar* and *pie charts* are typically covered in elementary statistics texts (see, for example, [12], [40], [77], or [95]), and articles of studies conducted on the effectiveness of using such charts appear in the literature (see, for example, [121], [125], and [137]).

The data frame `deerCounts` is used for earlier illustrations in this chapter. This data frame contains a random sample of 175 rows of the factors `month`, `gender`, `island`, and `weather` from the previously seen data frame `deerStudy`. Open this chapter's script file to follow along and see in it additional explanations of the code presented.

7.2 Bar Charts

In *bar charts* the frequencies corresponding to factor levels are represented by lengths of bars having equal width. The orientation (vertical or horizontal) used for bar charts is typically vertical, but varies from author to author.[1] In R, the default orientation of the plot produced by the function `barplot` is vertical. The horizontal axis represents factor levels, and the vertical axis measures the height of the bars in terms of frequencies. Bar charts provide an effective means of comparing frequencies across factor levels.

The usage definition for the function `barplot`, with arguments of interest, is

```
barplot(height, width = 1, space = NULL,
    horiz = FALSE, ...)
```

where `height` is a vector or a matrix of frequencies, `width` is an optional vector of bar widths, `space` is used to specify the amount of space between bars, `horiz` specifies whether the bars are to be plotted horizontally or vertically (default), and "..." denotes a fairly long list of additional arguments and graphical parameters that can be passed into this function. Illustrations of the use of some such arguments and parameters appear in the following examples.

For One-Way Tables

A frequency distribution of deer counts by month can be used to construct a default bar chart using, for example,

```
> (nObs <- with(data = deerCounts, expr = table(month)))
> barplot(height = nObs)
```

To construct a horizontally aligned bar chart, include the argument assignment `horiz = TRUE` in this function call. The arguments `width` and `space` can be used in a variety of ways to alter the appearance of the plot, including replacing the bars by vertical lines (see the script file).

As for most plotting function calls in R, a default sized (7×7 in) graphics device window is automatically opened. Alternatively, the function `win.graph` can be used to prepare a custom-sized graphics device window, as well as set the pointsize to be used for plotting images in the figure.[2] The function `win.graph` is one of a group of functions that can be used to open a graphics device window with some basic formatting in place, see Section 7.6.1 for other options including those suitable for non-Windows platforms. From here on, most graphics windows in this and later chapters are sized using the function `win.graph` and this function call will typically appear only in the script files.

The construction of a bar chart from `nObs` provides an opportunity to illustrate some basic graphics formatting capabilities of R. For example, Figure 7.1 is the result of running the following code. Begin by opening a custom sized window, then construct a vertical bar chart of the frequencies in `nObs`.

---

[1] In [119, pp. 37–64], for example, a chart with horizontally aligned bars is called a bar chart, and one in which the bars are vertically aligned is called a *column chart*.

[2] Macintosh users should use the function `quartz` in place of the function `win.graph`.

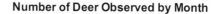

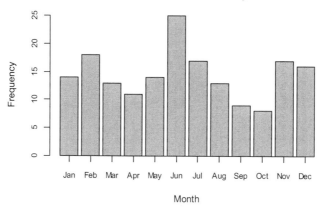

FIGURE 7.1

Bar chart of frequency of deer observed during each month of the year.

```
> win.graph(width = 5, height = 3.5, pointsize = 10)
> barplot(height = nObs, horiz = FALSE,
+    xlab = "Month", ylab = "Frequency"
+    ylim = c(0, 25),
+    cex.axis = 0.75, cex.names = 0.75,
+    names.arg = month.abb, axis.lty = 1,
+    main = "Number of Deer Observed by Month")
```

The various argument assignments in this function call include: `ylim`, an optional argument for the range of values for the vertical axis; `xlab`, the label for the horizontal axis; `ylab`, the label for the vertical axis; and `main`, the figure title. The assignment `names.arg = month.abb` makes use of the built-in constant `month.abb` to assign abbreviated month names to each bar, and if `axis.lty = 0` (the default setting) is used, a horizontal axis is not drawn. See the script file for line-by-line documentation/comments.

Pareto Charts

If, for a given frequency distribution, the levels of the factor are of nominal scale, the reader's attention can be drawn to the most dominant levels by arranging the bars in descending heights. Such bar charts are called *Pareto charts* and can be obtained with the help of the `sort` function. For example,

```
> clouds <- with(data = deerCounts,
+     expr = sort(x = table(weather), decreasing = TRUE))
> barplot(height = clouds, col = c("green", "blue", "red"))
```

produces a Pareto chart (figure not shown). The optional `col` argument enables the creation of custom colored bars, further color options can be found in the `palette` documentation page.

For Two-Way Tables

Consider, again, constructing a bar chart of deer counts by month, but one in which frequencies are distributed among two bars per month: one for males and one for females. Once again, first create a contingency table, and then construct the bar chart shown in Figure 7.2 using

```
> (nObs <- with(data = deerCounts,
+     expr = table(gender, month)))
> barplot(height = nObs, beside = TRUE, space = c(0, 0.5),
+     xlab = "Month", ylab = "Frequency",
+     cex.axis = 0.75, cex.names = .75,
+     axis.lty = 1, legend.text = c("Male", "Female"),
+     args.legend = list(bty = "n", cex = 0.75),
+     names.arg = month.abb, cex.main = 1.15,
+     main = "Frequency of Deer Observed
+     by
+     Month and Gender")
```

The splitting of the figure title in Figure 7.2 is accomplished by splitting the character string assigned to the `main` argument into the desired number of lines. Note that the indentation of the new line in this part of the code must be the same as the previous line; see what happens if the indentation differs. This "new line" capability can also be applied to axis labels in `xlab` and `ylab`.[3]

The assignment `beside = TRUE` gets the bars placed side-by-side, and the assignment `space = c(0, 0.5)` sets the space between bars as a percentage of average bar-width (0 for within a month, and 0.5 for between months).

The assignment `legend.text = c("Male", "Female")` identifies the gender levels,[4] the assignments to `args.legend` suppress the plotting of a box

---

[3]The same may be accomplished by including an appropriately placed "\n" in character strings assigned to these arguments directly, or by way of the `paste` function. For additional text formatting options, see the `Quotes` documentation page.

[4]It is important that the legend text have the same order as they appear in the rows of the contingency table.

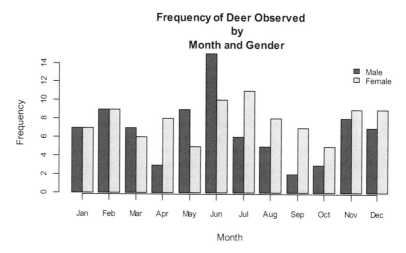

FIGURE 7.2
Bar chart of frequency of deer observed by gender for each month of the year.

about the legend (`bty = "n"`) and adjusts the legend size (`cex = 0.75`). If so desired, the argument `col` can be used to reset bar colors to a preferred combination. See Section 7.6.5 for more on legends.

For Three-Way Tables

Consider obtaining a bar chart that portrays observed frequencies by `gender`, `month`, and `island`. The code

```
> (nObs <- with(data = deerCounts,
+    expr = table(gender, month, island)))
```

constructs and displays the relevant 3-dimensional array of frequencies that contains contingency tables for each of the four islands. The largest frequency in this array is 6, and the (dimension) names of the four tables can be extracted and placed in a vector `islands` as shown in the script file.

Bar charts for the four contingency tables in `nObs` can be placed in a single suitably sized window (for example, 8 × 7 in.) that is partitioned into four equal-sized compartments by running

```
> par(mfrow = c(2, 2), ps = 10)
```

right after the `win.graph` (or `quartz`) function call to prepare the window. The assignment `mfrow = c(2, 2)` in the function `par` partitions the window into compartments of two rows and two columns, and `ps = 10` sets the point-size to 10.[5]

Now a for-loop can be used to construct the four two-way bar charts as follows. First get the index vector

```
> islands <- dimnames(nObs)[[3]]
```

Then the following code places the four bar charts in the compartments by rows, in the order in which they are constructed (figure not shown).

```
> for (i in islands)
+ {    # Start loop body
+       barplot(height = nObs[ , , i], ylab = "Frequency",
+           ylim = c(0, 6),
+           beside = TRUE, space = c(0, 0.5),
+           cex.axis = 0.75, cex.names = .75,
+           axis.lty = 1, legend.text = c("Male", "Female"),
+           args.legend = list(bty = "n", cex = 0.75,
+               x = "topleft"),
+           xlab = "Month", names.arg = month.abb,
+           cex.main = 1.15, main = i)
+ }    # End loop body
```

See the script file for sample code in which, with the help of some additional R graphics features and functions, the `apply` function is instead of the for-loop to duplicate the figure.

7.3 Dot Charts

The *dot chart* is an alternative to the bar chart, and displays frequencies as dots instead of bars. In figures produced by the function `dotchart`, the vertical axis represents the categories being displayed and the positions of the dots in the horizontal scale indicate the frequencies of the categories. The usage definition for this function, including arguments of interest, is

---

[5] An equivalent argument assignment in the function `par` is `mfcol = c(2, 2)`. The same partitioning is accomplished; however, in this case the bar charts are placed into their respective compartments by columns instead of by rows. See Section 10.1 for alternatives to the `mfrow` and `mfcol`.

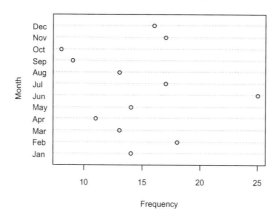

FIGURE 7.3
Dot chart of deer observed by month. This chart is equivalent to a horizontally aligned bar chart.

```
dotchart(x, labels = NULL,
    main = NULL, xlab = NULL, ylab = NULL, ...)
```

Additional arguments not listed are left in their default settings for the following examples.

For One-Way Tables

The dot chart equivalent of Figure 7.1 appears in Figure 7.3. First reconstruct the needed frequency distribution

```
> nObs <- with(data = deerCounts, expr = table(month))
```

Then, after opening a suitably sized window with desired parameter settings, Figure 7.3 is produced by running

```
> dotchart(x = as.vector(nObs), labels = month.abb,
+    xlab = "Frequency", ylab = "Month",
+    main = "Number of Deer Observed by Month")
```

Since the `dotchart` function uses a vector or a matrix in its construction of a dot chart, the function `as.vector` is used to convert `nObs` in this function call.[6]

---

[6]This avoids a warning message being printed to the console. For this case, if this is not done the `dotchart` function converts the table `nObs` into a numeric vector using the function `as.numeric`.

One might wish to improve the readability of a plot such as Figure 7.3 by including line segments to represent frequencies. Running the code

```
> segments(x0 = 0, y0 = 1:12, x1 = nObs, y1 = 1:12)
```

after the above function call accomplishes this task.

For Two-Way Tables

To construct Figure 7.4, the dot chart equivalent of Figure 7.2, first obtain the contingency table for month by gender,

```
> nObs <- with(data = deerCounts,
+     expr = table(month, gender))
```

Then, using a suitably sized window, run

```
> dotchart(x = nObs, labels = month.abb,
+     xlab = "Frequency",
+     main = paste("Frequency of Deer Observed\n",
+         "by\nMonth and Gender"))
```

to obtain the dot chart.[7] Finally, Figure 7.4 is completed by inserting the factor name "Month" in the left-hand margin using

```
> mtext(text = "Month", side = 2, cex = .75, line = 2)
```

The function `mtext` permits the plotting of a given string of `text` on a desired `side`, and in a chosen margin `line`. See Section 7.6.6 for more on this function.

Again, if so desired horizontal lines can be included in the plot with the help of the `segments` function.

Note that for this figure the factor `gender` appears second in the `table` function call. If the order of `gender` and `month` is switched, twelve vertically arranged dot charts are constructed.

For Three-Way Tables

As with bar charts, three-way tables can be represented graphically by a sequence of dot charts, one chart for each level of the third factor. However, when working with such dot charts it is difficult to place all figures in a single suitably sized graphics window which will fit on a page, and also retain clarity. This being said, as shown in the script file, code to produce three-way dot chart equivalents of the bar charts constructed in Section 7.2 can be easily prepared; see the script file for the code and details.

---

[7]Note that the conversion of the object `table(month, gender)` to a matrix is not required here. If it were required, code of the form `as.matrix(nObs)` would be used.

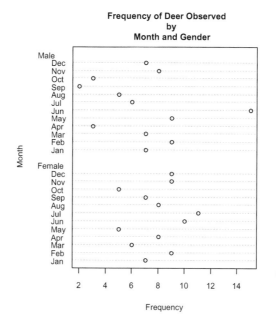

FIGURE 7.4
Dot chart of frequencies of deer observed by month and gender.

7.4 Pie Charts

It is suggested that pie charts are best used when estimating the size of a proportion (of the whole) enters into comparisons of frequencies across categories (see, for example, [125] and [137]). It is advised, however, that the number of categories be at most five (see, for example, [119]).

The usage definition for the `pie` function along with arguments of interest is

```
pie(x, labels = names(x), radius = 0.8, col = NULL,
    main = NULL, ...)
```

Illustrations of some of the ways in which each argument can be used appear in the following examples.

For One Factor

Pie charts portray frequencies of the levels of a factor as percentages of the whole. So, the first task is to obtain a relative frequency table for the levels of `island`.

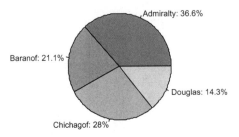

FIGURE 7.5
Pie chart of percentages of deer observed by islands.

```
> nObs <- with(data = deerCounts, expr = table(island))
> (rFreq <- round(prop.table(x = nObs), digits = 3))
```

Labels for each sector, including the percentages, can be prepared in advance using the **paste** function,

```
> (secNames <- paste(names(rFreq),": ",
+       rFreq*100, "%", sep = ""))
```

The **paste** function concatenates given collections of vectors into a vector of character strings. Here the island names are concatenated with their corresponding percentage. See Section 7.6.3 for more on the **paste** function.

After preparing a suitably sized window to include all user-inserted labels completely (often a trial and error process), Figure 7.5 is obtained by running

```
> pie(x = rFreq, radius = 1, labels = secNames,
+       col = gray(seq(from = 0.5, by = 0.125,
+             length.out = 4)),
+       main = "Percents of Observations by Island")
```

As with the **barplot** function, a wide range of color schemes can be used for the sectors through the **col** argument.

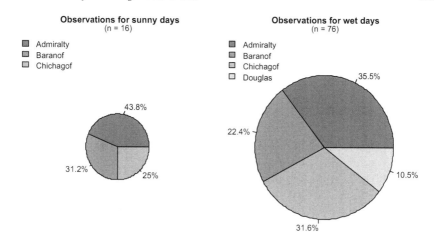

FIGURE 7.6
Pie charts for deer observed by islands, for sunny and for wet days. The difference in sample sizes for the two types of days are reflected in the sizes of the two charts.

For Two Factors

Consider constructing two side-by-side pie charts of deer observations by islands on sunny days and wet days only, such as shown in Figure 7.6.

Begin by obtaining the contingency tables for the two types of days of interest (sunny and wet) by restricting the extraction of frequencies to the weather factor levels of interest,

```
> (nObs <- with(data = deerCounts,
+     expr = table(weather, island))[2:3,])
```

and then obtain a relative frequency table of observations across islands for each type of day rounded to three decimal places.

```
> rFreq <- round(prop.table(x = nObs, margin = 1),
        digits = 3)
```

In addition to the fact that no deer were observed on Douglas during sunny days, running

```
> (nSamp <- rowSums(nObs))
```

shows fairly large differences in the sample sizes by levels of **weather**. This fact can be reflected in the resulting pie charts in two ways. One is to insert the total frequency for each chart, and another is to adjust the areas of the pie charts accordingly to alert readers to differences in sample sizes. Since the radius of a circle is proportional to the square root of the area of the circle, the radii of the pies can be set using

```
> rad <- sqrt(nSamp/sum(nSamp))*0.8
```

Note that the default radius is 0.8.

Next, gray scale sector colors and chart titles, and labels containing the sample sizes for each type of day can be prepared in advance using the `paste` function.

```
> sectorColors <- gray(seq(from = 0.5, by = 0.125,
+     length.out = 4))
> titles <- paste("Observations for",
+     row.names(nObs), "days")
> nLabel <-paste("(n = ",nSamp,")", sep = "")
```

Using a desired sized window (for example, 6 × 3.5 in.), the graphics margin parameter can be altered so that the plotting window accomodates both figures and maximizes the plotting area. For example, by running

```
> par(mfrow = c(1, 2), ps = 10, cex = 0.75,
+     mar = c(0, 0, 2, 0))
```

the number of lines assigned to the margin of each side of the plotting screens are altered through the `mar` argument – all sides except the top have no margin lines, and the top has two.[8]

Finally, to construct the two pie charts as shown in Figure 7.6, run

```
> for (i in 1:2)
+ {   # Begin loop body
+     p <- rFreq[i, ]*100
+     pie(x = p[p > 0], radius = rad[i],
+         col = sectorColors[p > 0],
+         labels = paste(p[p > 0],"%",sep = ""))
+     legend(x = "topleft", horiz = FALSE, bty = "n",
+         legend = names(p)[p > 0],
+         fill = sectorColors[p > 0])
+     title(main = titles[i], line = 1)
```

[8] The sides in `mar` are assigned in the order bottom, left, top, and right. The default margin setting of a plotting window is `mar = c(5, 4, 4, 2) + 0.1`.

```
+       mtext(side = 3, line = 0, cex = .75,

+           text = nLabel[i])

+ }   # End loop body
```

Note the use of the `legend`, `title`, and `mtext` functions. These functions permit after-the-fact additions to a graphics image. See Sections 7.6.4–7.6.6 for more on these functions.

For Three Factors

Consider constructing pie charts for the four islands, for cloudy and wet days for each gender. This provides an opportunity to use nested for-loops in a graphics construction setting. Here is an outline of one way in which this approach can be used to construct such a figure (see the script file for details).

Begin by constructing the three-way frequency distribution, excluding data for sunny days. The result will be a 3-dimensional array, the dimension names of which can be determined and used to prepare a matrix of titles (named `titles`) for the four charts. Next, the matrix of sample sizes for the various combinations of weather and gender is obtained and used to construct a matrix of sample size labels (named `nLabels`) for each chart. These labels will be placed below each chart title. The relative frequencies for observations within each weather-gender combination are then computed (named `rFreq`) and these are used to determine the angles making up the slices within the pies. Finally, the sample size dependent radii for the pies (named `rad`) to be used for the pies are obtained, and the slice colors (named `colors`, in gray scale) are predefined for each island.

With this information and a suitably sized and formatted 2×2 plotting window, the pie charts can then be constructed using nested for-loops, one for weather and the other for gender.

7.5   Exporting Graphics Images

The images contained in this book that were constructed in R were copied as *metafiles* directly from the graphics device window.[9] However, there are a variety of other options available that may be preferred for specific purposes.

A convenient (and versatile) approach to exporting graphics images through code is to use the `savePlot` function, which saves the current plot

---

[9]In the Windows environment, right click the mouse on the graphics device window, then select **Copy as metafile**.

(in the active graphics device window) to a file. The usage definition, with arguments of interest, for this function has the form

```
savePlot(filename, type)
```

where `type` is assigned any one of `"wmf"` or `"emf"`, `"png"`, `"jpeg"` or `"jpg"`, `"bmp"`, `"tiff"` or `"tif"`, `"eps"` or `"ps"`, and `"pdf"`. The `filename` argument can be assigned a file name (and path, as appropriate), alternatively the assignment `filename = file.choose()` can be used[10] as in the case of the `save` and `load` functions. All other arguments listed in the usage definition are left in default setting in the following illustration.

Other functions that can be used to export graphics images obtained in R directly through code are listed in the following table.

To create file of type	Use the function
Windows Metafile (*.wmf)	win.metafile
Bitmap (*.bmp)	bmp
JPEG (*.jpeg)	jpeg
TIFF (*.tiff)	tiff
PNG (*.png)	png
PDF (*.pdf)	pdf
Postscript (*.ps)	postscript

Here are a couple of examples.

Using `savePlot` to Export a Graphics Image to a *.eps File

Here, use the `savePlot` function to export Figure 7.1 to an *encapsulated postscript* (*.eps) file. First recreate Figure 7.1 (see script file), then run

```
> savePlot(file = file.choose(), type = "eps")
```

The process, then, is the same as for the `save` function; in the **Select file** window, go to the desired folder and then open (or create) the file to which the image is to be exported. Once an image has been exported to a file (of the desired type), it can be imported/inserted into a document (and resized/rescaled as needed) using the appropriate steps.

Using `pdf` to Export a Graphics Image to a *.pdf File

A graphics image, say of Figure 7.6, can be exported to a *.pdf file using the `pdf` function in much the same manner as was the case for the `savePlot` function except in that here the process is begun by first opening (or creating) the file to which the image is to be saved by running

---

[10]Macintosh users may find that the `file.choose()` assignment will not work; see comment at the start of the script file.

```
> pdf(file = file.choose(), width = 6, height = 3.5)
```

Next, run the image construction code (see the script file). Finally, complete the process by running

```
> dev.off()
```

The exported image can then be imported/inserted into a document and re-sized/rescaled as appropriate.

The general implementations of the other functions listed above are the same. First identify the file/location to which the image is to be exported using the appropriate function, then create the image, and finally run dev.off().

7.6   Additional Notes

This section looks at a selection of graphics capabilities available for customizing plotting windows, further examples of using the **paste** function to prepare character strings to insert in figures, more on the **title**, **legend** and **mtext** functions, and some examples illustrating the capabilities of the **text** function.

7.6.1   Customizing Plotting Windows

The **win.graph** function, which includes only three arguments, **width**, **height**, and **pointsize**, along with desired graphics parameter settings made through the **par** function suffices for customizing most plots in a Windows environment. Alternatives to the **win.graph** function for Windows operating systems include the functions **windows** and **x11** (**x11** works in Linux and Unix). For the OS X operating system (Macintosh) the function **quartz** is used to start up the graphics device driver.

In the Windows environment the **windows** function provides the most additional features. For example, running the code

```
> windows(width = 5, height = 3.5, pointsize = 10,
+     bg = "dark green", family = "serif")
> barplot(height = nObs, horiz = FALSE,
+     xlab = "Month", ylab = "Frequency",
+     ylim = c(0, 25), col = "gold",
+     cex.axis = 0.75, cex.names = .75,
+     names.arg = month.abb, axis.lty = 1,
+     main = "Number of Deer Observed by Month")
```

duplicates Figure 7.1 in all respects except in that the background color is dark green (`bg = "dark green"`) and the the font used for the text in the figure is `family = "serif"`. The color of the bars is altered using `col = "gold"` in the `barplot` function.

The difference between the `windows` function (as used above) and the `x11` function is that the font family cannot be set within `x11`. This, however, can be accomplished using the `par` function, for example, if

```
> x11(width = 5, height = 3.5, pointsize = 10,
+     bg = "dark green"); par(family = "serif")
```

is used in place of the previous `windows` function call before the `barplot` function call, the result is an identical plot. Alternatively, as for most plotting functions, the font family parameter can be set locally within the `barplot` function call.

In the OS X operating system, figures obtained in the previous examples with `windows` and `x11` can be duplicated by running

```
quartz(width = 5, height = 3.5, pointsize = 10,
       bg = "dark green", family = "serif")
```

before the previously run `barplot` function.

7.6.2   The `plot.new` and `plot.window` Functions

Illustrations in the remaining sections of this chapter make use of these two functions, their usage definitions being

```
plot.new()
plot.window(xlim, ylim)
```

These functions are useful when there is a need to move from one plotting frame to another in a partitioned graphics window, or prepare figures from scratch using low-level plotting functions.[11]

The `plot.new` function serves to create, or advance to, a new plotting frame. The `plot.window` function can be used to set up the coordinate system for a plotting frame opened by `plot.new` through the arguments `xlim` (the horizontal axis range), and `ylim` (the vertical axis range). Examples of using these functions are contained in the code for this and the remaining sections of this chapter.[12]

---

[11]For example, `legend`, `mtext`, `title`, and several others that can be run only after a high-level plotting function call, such as a `barplot` or `pie`. Later chapters cover a variety of additional low-level plotting functions.

[12]See, also, the script file code for Section 7.2 where the `apply` function is used to construct bar charts across levels of three factors.

7.6.3 More on the `paste` Function

One of the uses of this function, as seen so far, is in preparing character strings for insertion into a graphics image using the `title`, `legend`, `mtext`, or `text` functions. Manipulations of two components within a `paste` function call are demonstrated here, the vectors to be concatenated and the `sep` argument which specifies how the concatenated characters are to be separated.

At the simplest level, the purpose of the `sep` argument (which defaults to a single space, `sep = " "`) can be demonstrated by running each of the following

```
> paste("A", 1)
> paste("A", 1, sep = "")
> paste("A", 1, sep = "-")
```

Similarly,

```
> paste(c("A", "B", "C"), 1, sep = "")
```

shows that when a vector of length 1 is concatenated with a vector of length greater than 1, its entries are recycled (much as for arithmetic operators; see Section 6.1). Similarly, when two (or more) vectors of equal length are passed into the `paste` function the concatenations are performed termwise.

```
> paste(c("A", "B", "C"), 1:3, sep = "")
```

The recycling feature involving vectors can often be taken advantage of in a variety of ways. See the script file for further examples, including one in which the result of a calculation is concatenated with character strings.

7.6.4 The `title` Function

This function can be used to insert various labels and titles in an existing plot. The usage definition with basic arguments is

```
title(main = NULL, sub = NULL, xlab = NULL, ylab = NULL,
      line = NA, ...)
```

The arguments `main`, `sub`, `xlab`, and `ylab` are used to pass in a main and subtitle for a plot, as well as labels for the x- and y-axes. The `line` argument can be used to change the default line (in the margins) where the various titles or labels are placed. Additional graphics parameters, for example, to control the colors (`col.main`, `col.sub`, etc), scaling (`cex.main`, `cex.sub`, etc), position adjustments (`adj.main`, `adj.sub`, etc), and others can also be passed into the `title` function. See the script file for sample exploratory code.

7.6.5 More on the `legend` Function

This function was used in Section 7.4 at a basic level in the construction of pie charts. The general usage definition for the `legend` function, along with a small selection of arguments is

```
legend(x, y = NULL, legend, fill = NULL, bty = "o",
    horiz = FALSE, xjust = 0, yjust = 1)
```

Here are explanations of the arguments listed; run the code in the script for this section to follow this discussion.

The `fill` argument is used to color boxes corresponding to the legends assigned to `legend`. This can be a vector of colors or a vector of integers (for example, `"black"` or 1, `"red"` or 2, ...). The default coloring is `"transparent"`, or 0.

The location where the legend is inserted can be specified by assigning any one of the character strings `"bottomright"`, `"bottom"`, `"bottomleft"`, `"left"`, `"topleft"`, `"top"`, `"topright"`, `"right"` and `"center"` to the argument x. Alternatively, the location of insertion can be specified using coordinates assigned to x in the horizontal scale and y in the vertical.

Legends can also be oriented horizontally (`horiz = TRUE`) or vertically (`horiz = FALSE`), and a box around the legend can be included (`bty = "o"`) or suppressed (`bty = "n"`).

When the arguments x and y are used to identify the point of insertion by coordinates, the `xjust` and `yjust` can be used to determine how the legend text is justified about the point of insertion. If `xjust = 0` (default), the legend is placed to the right of the vertical line at x, and if `xjust = 1` the legend is placed to the left of the vertical line at x. If `yjust = 1` (default), the legend is placed below the horizontal line at y, and if `yjust = 0` the legend is placed above the horizontal line at y.

A variety of symbols (through the `pch` argument) or line types (using the `lty` argument), instead of the colored boxes seen so far, can also be plotted with legends in mathematical annotation. Examples of such applications appear in Chapter 8, and see Section 9.1.2 for further details on the `pch` parameter.

7.6.6 More on the `mtext` Function

The general usage definition for the `mtext` function, including only those arguments of interest here, is

```
mtext(text, side = 3, line = 0,
    at = NA, adj = NA, padj = NA)
```

Here are explanations for each of the listed arguments, run the sample code in the script file to get figures to help to follow the discussions below. Begin with the arguments `text`, `side`, and `line`.

At the simplest level, the text assigned to `text` is centered on a specified margin `line` and `side` of the plotting window. The sides are identified by 1 (bottom), 2 (left), 3 (top, the default side), and 4 (right). The margin lines are typically integers selected from 0, 1, 2, ... to the maximum number of margin lines that are defined for the side in question.[13] However, negative integers can also be used.

Multiple margin insertions can be performed simultaneously if the `text` argument is assigned a vector containing more than one entry. In this case each component of this vector can be assigned its own line through the `line` argument, and its own side through the `side` argument.

The `at` argument can be used to place the inserted text at any position on a specified line (and side).

The `adj` and `padj` functions provide control on how the inserted text is positioned with respect to the location at which it is inserted. Under the default settings `adj` centers the text horizontally, and `padj` places the text at the top of the line in which it is inserted.

If `padj` is left unaltered, assigning values starting with 0 and ending with 1 to `adj` alters the horizontal position on sides 1 and 3 from right to left with respect to the position of insertion, and the vertical position on sides 2 and 4 from top to bottom.

If `adj` is left unaltered, then assigning values starting with 0 and ending with 1 to `padj` alters the vertical position on sides 1 and 3 from top to bottom, and the horizontal position on sides 2 and 4 from left to right.

Combinations of assignments to `adj` and `padj` result in combinations of positioning with respect to the location of insertion.

7.6.7 The `text` Function

The general usage definition for the `text` function, with four arguments of interest, is

```
text(x, y = NULL, labels, adj = NULL)
```

For the following discussion it is assumed that both the `x` and `y` arguments are used. Run the code for this section in the script file to follow along.

Begin with plotting a single character string. Values assigned to the arguments `x` and `y` identify the coordinates in the rectangular plane where the character string assigned to `labels` is plotted. The argument `adj` is used to adjust the location of the plotted text relative to the position of insertion.

If the `adj` argument is not included in the `text` function call (`adj = NULL`), then the plotted text is centered (horizontally and vertically) on the position (x, y) specified in `x` and `y`. There are two ways in which the `adj` argument can be used.

---

[13]In the default setting, counting `line = 0` as the first line, these are 5 for the bottom side, 4 for the left and top, and 2 for the right.

If an assignment of the form `adj` = a is used, where $0 \leq a \leq 1$, then the plotted text is centered about the vertical position y, and moves from the right of the horizontal position x (left-justified) when $a = 0$ to the left of the horizontal position x (right-justified) when $a = 1$.

An assignment of the form `adj` = `c(a, b)` can be used, with $0 \leq a, b \leq 1$, to adjust both the horizontal and vertical positioning of the plotted text with respect to the point of insertion, (x, y). In this case a controls the horizontal position as described previously. The vertical positioning moves from above the vertical position y when $b = 0$ to below the vertical position y when $b = 1$.

An assignment of `adj` = `c(0.5, 0.5)` results in the same positioning as when the argument `adj` is left out of a text function call; this is the default positioning. If `labels` is assigned a vector of character strings for insertions, then at least one of `x` or `y` should be assigned an equal length vector of positions for the character strings in `labels` to be inserted at the positions identified by `x` and `y`.

8

Basic Plots for Numeric Data

This chapter covers some common graphical methods for numeric data addressing exploratory tools used to assess distributional properties of data. Like the previous chapter, the methods described here provide graphical representations of frequency distributions and data summaries. Open the script file, `Chapter8Script.R`, and run the code to follow along.

8.1 Histograms

Grouped frequency distributions can be displayed graphically using a *histogram*. For example, in Section 5.2, a grouped frequency distribution of altitudes at which deer were observed was obtained. Figure 8.1 shows the histogram constructed from this grouped frequency distribution. This section begins with histograms obtained using default settings in the `hist` function, and ends with a selection of customized histograms, Figures 8.1 through 8.3. The usage definition for the function `hist` with a selection of arguments that are of interest here is

```
hist(x, breaks = "Sturges", freq = NULL,
     include.lowest = TRUE, right = TRUE,
     main = paste("Histogram of" , xname), sub = NULL,
     xlim = range(breaks), ylim = NULL,
     xlab = xname, ylab, axes = TRUE, labels = FALSE, ...)
```

For convenience, extract the vector `altitude` from `deerStudy`, then a default plot at the simplest level can be obtained by running (figure not shown)

```
> hist(x = altitude)
```

The axes labels and the main title can be changed by assigning desired character strings to the arguments `xlab`, `ylab`, and `main`, and a subtitle can be inserted by including the `sub` argument in the function call.

While default histograms are adequate for a quick examination of the data, a fine-tuning of the appearance is sometimes desirable. The following discussions illustrate some techniques that can be employed for this purpose.

Setting Class Boundaries

In the default setting the class boundaries (or breakpoints) are determined by an algorithm which uses the number of classes, k, suggested by Sturges' formula (breaks = "Sturges"),

$$k = \lceil \log_2(n) + 1 \rceil,$$

where n is the sample size.[1]

There are two other options for algorithms that can be used to determine the class boundaries for a histogram: One uses the number of classes suggested by *Scott's formula* (breaks = "Scott"),

$$k = \left\lceil \frac{[\max(x) - \min(x)] \sqrt[3]{n}}{3.5\,s} \right\rceil,$$

s being the sample standard deviation; the other uses the *Freedman–Diaconis formula* (breaks = "FD"),

$$k = \left\lceil \frac{[\max(x) - \min(x)] \sqrt[3]{n}}{2\,\text{IQR}} \right\rceil,$$

IQR being the interquartile range of the sample. See the script file for some exploratory code. These seem to indicate that the hist function is designed to take a suggested number of classes, then use this number as a starting point in computing boundary points occurring at "nice" values.

Another way to establish class boundaries and the number of classes is to assign the argument breaks a vector of user-determined class boundary points (or a function call that generates such a vector). For example, the maximum altitude is exactly 4200 feet, and suppose the number of classes desired is 8, then running

```
> hist(x = altitude,
+      breaks = seq(from = 0, to = 4200, length.out = 9),
+      include.lowest = TRUE, right = FALSE)
```

produces a histogram with 8 bins. The assignments to the arguments include.lowest and right mean the same as for the cut function used to construct the grouped frequency distribution in Section 5.1. There will be occasions, as with this histogram, when the axes tickmarks do not fall on the class boundaries. No worries! See Figure 8.1 and read on.

Customizing the Axes

By suppressing the plotting of axes in the hist function call (use axes = FALSE), it is possible to plot customized axes on any side through the use of

---

[1]Note that the notation $\lceil \cdot \rceil$ represents the ceiling function, ceiling in R.

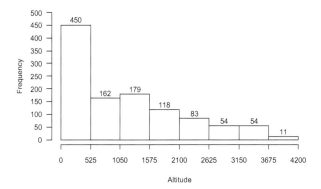

FIGURE 8.1

An example of a frequency histogram with customized class boundaries, axes scaling and tickmarks, and bin labeling.

the function `axis`. If this is to be done, it is also useful to specify axes limits through the arguments `xlim` and `ylim` in `hist`.

For example, including specified axes ranges, axes labels and a main title, the following code produces a histogram with no axes.

```
> hist(x = altitude,
+      xlim = c(0, 4200), ylim = c(0, 500),
+      xlab = "Altitude",
+      main = "Altitudes at which Deer were Observed",
+      breaks = seq(from = 0, to = 4200, length.out = 9),
+      include.lowest = TRUE, right = FALSE,
+      axes = FALSE, labels = TRUE)
```

To include a customized horizontal axis below the histogram (`side = 1`), and a vertical axis to the left of the histogram (`side = 2`), run

```
> axis(side = 1,
+      at = seq(from = 0, to = 4200, length.out = 9))
> axis(side = 2,
+      at = seq(from = 0, to = 500, by = 50), las = 1)
```

The result is Figure 8.1. The argument assignment `las = 1` in the second `axis` function call causes the tickmark labels in the vertical axis to be plotted horizontally. See Section 8.5.2 for more on the `axis` function, along with some related graphics parameter settings using arguments such as `las`.

Other Customizations

As with most plotting functions, additional arguments for resetting graphics parameters can be included in a `hist` function call. For example, see the result of running the code

```
> hist(x = altitude, main = NULL, xlab = "Altitude",
+      las = 1, col = "maroon",
+      cex.lab = 1.25, cex.axis = 0.75,
+      col.lab = "dark green", font.lab = 3,
+      col.axis = "dark red", font.axis = 2)
```

While such arguments are typically not listed in a plotting function's documentation page, they can be found in the `par` function's documentation page. The effects of altering such graphical parameters (from default settings) can then be explored by running code such as above.

Using Histograms to Compare Two or More Samples

Histograms can be used to perform empirical comparisons of the distributions of two or more populations; Figure 8.2 provides an example of this. By including the argument assignment `freq = FALSE` in a `hist` function call, the result is a density histogram (as opposed to the default frequency histogram where `freq = TRUE`). This figure also permits an opportunity to demonstrate another customization. On occasion, particularly with a sample having a large range, it may be convenient to perform an appropriate and easy transformation to convert the scaling of the data. Consider comparing the altitude preferences of male and female deer in the month of July.

For convenience, in this and later examples, extract data for only the month of July from `deerStudy` and then place altitudes for each gender, as separate vectors in a list.

```
> julyData <- subset(x = deerStudy,
+      subset = (month == "7"))
> genderAlts <- with(data = julyData,
+      expr = split(x = altitude, f = gender))
```

The ranges for the two samples can be examined using the `sapply` function, and based on the findings one might decide to construct the histograms over the altitude interval $[1000, 3000]$, which has a range of 2000 feet. By rescaling this interval into units of 200 feet each, the resulting interval becomes $[5, 15]$ which permits 10 classes of unit width. To perform this scaling, run the following code.

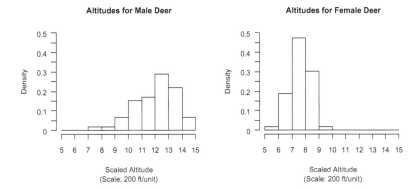

FIGURE 8.2
Density histograms for scaled altitudes at which deer of each gender were observed.

```
> scaledAlts <- sapply(X = genderAlts,
+       FUN = function(x){x/200})
```

Quick checks with exploratory runs of the `hist` function (see script file) on these scaled data suggest using a range of $[0, 0.5]$ for the densities of both genders. Finally, Figure 8.2 is produced on a 1×2 graphics window using a for-loop (see script file for code).

From these two histograms it can be seen that in July most male deer were observed at altitudes of around 2000 to 2800 feet, and females hung out at around 1200 to 1800 feet. Furthermore, while the sample of altitudes for males is skewed to the left, the sample for females appears approximately symmetric about the modal class.

Superimposing Density Curves on Histograms

By themselves histograms are useful tools for assessing symmetry in a sample. If a probability density curve is superimposed on a density histogram it is also possible to (approximately) assess what type of probability distribution the underlying random variable might (or might not) be.

For example, Figure 8.3 displays the normal density curve for the normally distributed random variable having the same mean and standard deviation as the sample `normal` (in `boxData`) superimposed on a density histogram for this sample. Here is an outline of how a figure such as Figure 8.3 may be constructed.

First, take a look at the range of values contained in the sample `normal` and settle on class boundaries, and then construct the corresponding density histogram. This may take a couple of trial runs to obtain an informative histogram.

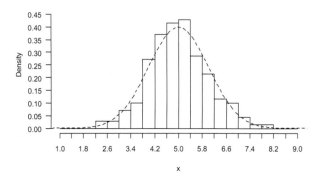

FIGURE 8.3
A density histogram of a sample with a superimposed normal density curve obtained using the sample mean and standard deviation.

Finally, to superimpose the normal density curve, two functions find use here: the function **dnorm** is used to obtain normal probability densities (see Section 11.1.3), and the function **curve** takes these densities and plots the corresponding curve. See the script file for code, and Section 9.3 for more details on the **curve** function.

Relative Frequency Histograms

The relative frequency equivalent of Figure 8.1 can be obtained quite easily in a roundabout way as follows. Using the same breakpoints and class interval point settings as for Figure 8.1, run

```
> relHist <- hist(x = altitude, plot = FALSE,
+     breaks = seq(from = 0, to = 4200, length.out = 9),
+     include.lowest = TRUE, right = FALSE)
```

Note that the assignment **plot = FALSE** suppresses the plotting of the histogram. Among the several objects contained in **relHist**, the object **counts** is of interest – this contains the frequencies for each class. Change the contents of **counts** to relative frequencies using

```
> relHist$counts <- relHist$counts/sum(relHist$counts)
```

Then, observing that the highest relative frequency is approximately 0.4, run

```
> plot(x = relHist, axes = FALSE,
+      xlim = c(0, 4200), ylim = c(0, 0.45),
+      xlab = "Altitude", ylab = "Relative Frequency")
```

to get the desired relative histogram without the axes.[2] Now, to wrap things up, the axes can be plotted using the approach used for Figure 8.1. The resulting figure (not shown) is a vertically scaled version of Figure 8.1 with bar heights representing relative frequencies as opposed to frequencies (or densities).

The same figure can be obtained by manipulating the contents of `density` in `relHist`; see the script file for sample code. See, also, Section 8.5.2 for an alternative method of displaying relative frequencies.

Cumulative Frequency Histograms

The trick used for relative frequencies in the previous example can be employed to construct cumulative frequency histograms as well, but in this case the frequencies in the histogram object are converted into cumulative frequencies. See the script file for sample code for constructing a cumulative frequency histogram for the `altitude` data.

8.2 Boxplots

Boxplots provide a graphical representation of the five-number summary introduced in Section 5.4.2. The R function for constructing boxplots is `boxplot`, the general usage definition of which, with arguments of interest, is

```
boxplot(x, horizontal = FALSE,
        range = 1.5, varwidth = FALSE, notch = FALSE, ...)
```

Here, the argument `x` accepts either a single numeric vector or a list (including a data frame) of numeric vectors. Boxplots can be plotted vertically (`horizontal = FALSE`) or horizontally (`horizontal = TRUE`) and, as with previous plotting functions the notation "..." represents additional argument assignments that can be used to alter a variety of default graphics parameter settings in a `boxplot` function call. Explanations and uses of the arguments `range`, `varwidth` and `notch` are provided in the following examples.

There are occasions when it is convenient to use an alternative definition of this function,

---

[2]The `plot` function is another of those versatile functions that "knows" what to do with an object of a particular class. In this case, `relHist` is of class "histogram."

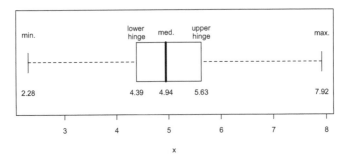

FIGURE 8.4
A basic boxplot of the sample `normal` from the list `boxData`. The inserted text identifying the various components and values in the five-number summary are plotted using the `text` function.

```
boxplot(formula, data = NULL, subset, ...)
```

In this case a data frame containing at least one numeric vector (say `y`), and one or more factors (say `fac1`, `fac2`, ...) is passed into the function via the argument `data`. The argument `formula` then accepts an expression of the form `y ~ fac1`, or `y ~ fac1 + fac2`, etc., depending on against how many factors boxplots of `y` are to be plotted. The `subset` argument permits a subsetting of the data if so desired. Additional arguments from the first definition can also be used in this second definition.

The following examples begin with two commonly seen boxplots and then illustrates two variations discussed in [90] and some other customizations.

Basic Plots

The boxplot in Figure 8.4 (excluding the additional text in the figure) is constructed using

```
> with(data = boxData,
+     expr = boxplot(x = normal, horizontal = TRUE,
+         range = 0, xlab = "Altitude"))
```

Setting `range = 0` causes the *whiskers* to (always) be extended all the way to the minimum value on the left, and the maximum value on the right. As mentioned in Section 5.4.2, the lower and upper hinges are close (and, under certain conditions, equal) in value to the first and third quartiles. In this type of boxplot the whiskers identify the ranges of the lower and upper 25% of the data values. In all boxplots the box always identifies the range of the middle 50% of data values (interquartile range)

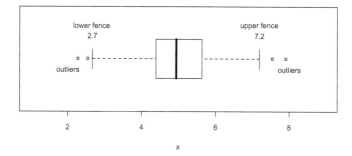

FIGURE 8.5
A modified boxplot of the sample `normal` from the list `boxData`. In this plot outliers are flagged using a cutoff of $1.5 \times$ IQR units from the lower and upper hinges.

In elementary statistics texts (for example, [12]) boxplots such as Figure 8.4 are referred to simply as boxplots, and are usually constructed using the "elementary" five-number summary discussed in Section 5.4.1.

Moderate and Extreme Outliers

In [12], boxplots in which outliers are identified are referred to as *modified boxplots*, and they differ from the previous type in that data satisfying certain user-specified criteria can be identified on the plot as being outliers, see Section 5.4. For example, in constructing Figure 8.5, setting `range = 1.5` in

```
> with(data = boxData,
+     expr = boxplot(x = normal, horizontal = TRUE,
+           range = 1.5, xlab = "x"))
```

causes the left whisker to be extended down to the *lower fence* (which is $1.5 \times$ IQR units below the lower hinge), and the right whisker to the *upper fence* (which is $1.5 \times$ IQR units above the upper hinge). The points plotted outside of these two fences are called *outliers*. If `range = 3` is used, the fences spread out further and any points plotted outside of these fences are called *extreme outliers*. Data that lie between $1.5 \times$ IQR and $3 \times$ IQR away from the lower and upper hinges are referred to as *moderate outliers*.

Customizing Appearances

The appearance of a boxplot can be refined quite a bit using various graphics arguments and functions already seen. For example, consider the construction of Figure 8.6.

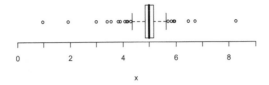

FIGURE 8.6
A customized boxplot of the sample `fat` from the list `boxData`. Here the plot frame and axis is suppressed, and a customized axis is included.

Begin by finding the range of the data in the vector `fat`, contained in the list `boxData`. Then, suppressing the plot-frame and the axes, construct the boxplot. The axis limits are set from 0 to 9, for the `boxplot` function this is always done through the `ylim` argument even if a horizontal boxplot is constructed.

```
> with(data = boxData,
+     expr = boxplot(x = fat, horizontal = TRUE,
+        range = 1.5, ylim = c(0, 9), xlab = "x",
+        axes = FALSE, pars = list(boxwex = 1.75)))
```

Note that the axis can be suppressed without suppressing the frame if `frame.plot = TRUE` is included in this function call.

Next, create vectors containing axis tickmark locations and labels.

```
> ticks <- 0:9
> tickLabels <- ifelse(test = (ticks%%2 == 0),
+     yes = ticks, no = "")
```

Note the use of the `ifelse` function to place blank labels at locations having odd integer values (see Section 8.5.1 for more on this function). Finally, plot the axis,

```
> axis(side = 1, at = ticks, labels = tickLabels)
```

and Figure 8.6 is completed.

Variable-Width Boxplots

When constructing boxplots over two or more factors levels (or samples), differences in sample sizes can be highlighted by varying the width of the plots constructed. This is done by including `varwidth = TRUE` in the `boxplot` function call. For example, Figure 8.7 is the result of the following code. First construct the boxplots, suppressing the axes and the frame.

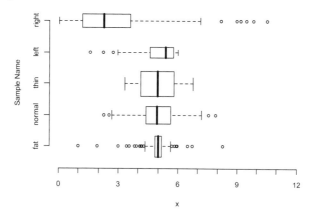

FIGURE 8.7
Variable-width boxplots of data contained in the list `boxData`. These boxplots also illustrate appearances of boxplots depending on the distributional nature of the samples plotted.

```
> boxplot(x = boxData, horizontal = TRUE,
+    range = 1.5, ylim = c(0, 12),
+    xlab = "x", ylab = "Sample Name",
+    axes = FALSE, varwidth = TRUE)
```

Next, create tickmark location and label vectors for the horizontal axis. in much the same manner as was done for Figure 8.6, and then plot the two axes.

Assessing Symmetry Through Boxplots

Plots such as Figure 8.7 can be used to assess symmetry in data. A longer right-whisker in a box plot indicates right-skewed data, and a longer left-whisker indicates left-skewed data. If the lengths of the whiskers are approximately equal, then a median to the left of the center of the box indicates right-skewed data, and a median to the right of the center of the box indicates left-skewed data.

If the whiskers are approximately equal in length, and if the median is approximately in the center of the box, then the sample is considered symmetric (at least approximately).

Notched Boxplots

A small alteration to a boxplot can be made by including `notch = TRUE` in a `boxplot` function call; see Figure 8.8 and the script file for sample code.

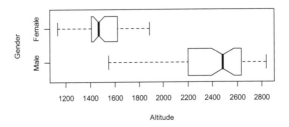

FIGURE 8.8
Variable-width boxplots with notches of the altitudes at which male and female deer were observed in July, from the data frame `julyData`. Here the horizontal axis is custom scaled, and the plots are enclosed in a frame.

In such boxplots the notches mark the 95% confidence interval

$$\left(\mathrm{med}(x) - 1.58\ \mathrm{IQR}\,/\sqrt{n},\ \mathrm{med}(x) + 1.58\ \mathrm{IQR}\,/\sqrt{n}\right).$$

For two samples that are close in size, if the notches do not overlap there is strong evidence to conclude that medians of the two populations differ.

Constructing Boxplots from Data Frames

Consider using boxplots to explore the spread of altitudes at which deer are sighted (from the `julyData` data frame) by gender and on each island. A quick plot can be constructed using the following function call (figure not shown).

```
> boxplot(formula = altitude ~ gender + island,
+    data = julyData, horizontal = FALSE)
```

There is, however, much to be desired with respect to the appearance of the resulting figure. An equivalent plot appears in Figure 8.12; see the code for that figure for ideas on one way to improve the appearance of the boxplot just constructed.

8.3   Stripcharts

While boxplots are very useful for assessing symmetry and the presence of outliers in data, they do not provide information on repeated values in the data and how data points are clustered relative to the whole; see Figure 8.9.

For small samples, *stripcharts* can be used to supplement information provided by a collection of boxplots such as Figure 8.7.

The general usage definition of the `stripchart` function, with arguments of interest, is

```
stripchart(x, method = "overplot", jitter = 0.1,
    vertical = FALSE, group.names, xlim = NULL,
    ylim = NULL, xlab = NULL, ylab = NULL, axes = TRUE,
    frame.plot = axes, add = FALSE, at = NULL, ...)
```

Discussions on the various forms that can be used for the argument x appear in the following examples. The options available for `method` are `"overplot"` (repeated values are over-plotted), `"stack"` (repeated values are stacked perpendicular to the numeric axis of measure), and `"jitter"` (*all* values are *randomly* jittered perpendicular to the numeric axis of measure). Of these three options, only `"jitter"` is of interest here. The argument `jitter` is used to set the amount of jittering desired and the argument `vertical` is used to orient the numeric axis vertically (`vertical = TRUE`) or horizontally (`vertical = FALSE`). If necessary, the argument `group.names` is used to assign names to the groups of data along the categorical axis. By default, these names are taken from the data. The `add` argument permits the user to either superimpose the stripchart on a current plot (`add = TRUE`), or create the stripchart in a new plotting window (`add = FALSE`). The `at` argument allows the user to assign specific locations along the categorical axis at which the data are plotted for each group. Finally, main and subtitles can be included in the usual ways, and the `axes` and `frame.plot` arguments are used to suppress or permit the plotting of axes and a frame for a chart.

Stripcharts as Alternatives to Boxplots

Consider displaying the samples in `boxData` using a jittered stripchart, such as Figure 8.9, *instead of* through boxplots as in Figure 8.7. After setting up an appropriate plotting window, the stripchart minus the plot frame and axes are constructed using

```
> stripchart(x = boxData, vertical = FALSE,
+    method = "jitter", jitter = 0.3, xlim = c(0, 12),
+    xlab = "x", ylab = "Sample Name", pch = 1,
+    group.names = names(boxData), axes = FALSE)
```

Then, code identical to that used for Figure 8.7 is used to include the axes for this plot. Note that the default value for `pch` is 0, and in this case the argument assignment `group.names = names(boxData)` is not needed (but was included for demonstration purposes). Using `vertical = TRUE`, and then altering axes

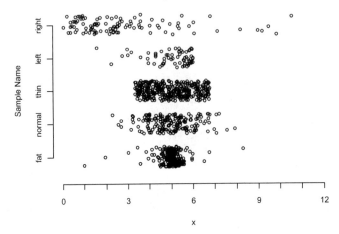

FIGURE 8.9
The stripchart equivalent of Figure 8.7. Observe that clusters of data are accentuated and the jittering of plotted points improves the visibility of repeated or approximately equal values.

constructions appropriately produces the vertically oriented equivalent of Figure 8.9.

The inclusion of appropriate horizontal (or vertical) lines representing cutoff values for the data permit another approach to performing a visual assessment for the presence of outliers in the data through stripcharts.

Superimposing Stripcharts on Boxplots

Consider combining the information in Figures 8.7 and 8.9 into a single plot; see, for example, Figure 8.10 and the script file for the code used.

For this plot, data within each group are separated into those that are outliers (plotted as asterisks) and those that are not (plotted as gray circles), and the fences for indicating outlier cutoff values are increased in length to improve clarity.

Stripcharts Across Levels of Two Factors Using a For-loop

The argument x can be assigned a formula of the form "y ~ `factor`"; this is particularly useful in looping structures, and also in obtaining plots across the levels of one or more factors without using a looping structure.

Here is an example of taking advantage of the formula method for the `stripchart` function within a for-loop. After running window partitioning and formatting code, stripcharts of altitudes for each of the four islands can be plotted by looping through the islands (see the script file for code). The

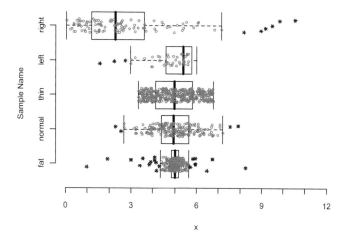

FIGURE 8.10
Here the boxplots from Figure 8.7 and the stripchart from Figure 8.9 are
superimposed on a single plot.

result is Figure 8.11, which shows a comparison of July altitudes (from the
earlier obtained data frame `julyData`) at which deer of each gender were
observed on each island.

Stripcharts Across Levels of Two Factors Without a For-loop

The formula method can be used to produce a single stripchart containing
plots equivalent to those in Figure 8.11. After preparing a suitable plotting
window, the `stripchart` function is used to plot the points with all labels,
axes and the plot-frame suppressed.

```
> stripchart(x = altitude ~ gender + island,
+     data = julyData, method = "jitter", jitter = .2,
+     pch = 1, vertical = TRUE, ylim = c(1000, 3000),
+     axes = FALSE, ylab = "", xlab = "",
+     group.names = NULL, frame.plot = FALSE,
+     at = 1:8)
```

Note the inclusion of the `at` argument in this function call. See the script file
for the remainder of the code. Note that Figure 8.12 does provide for a more
convenient comparison of the data across genders and islands than Figure
8.11.

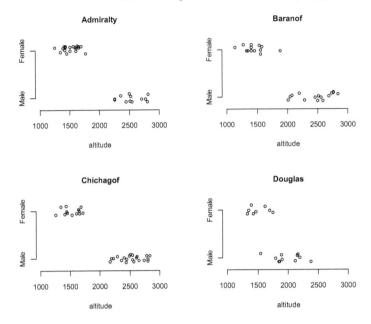

FIGURE 8.11
Stripcharts of July altitudes at which deer were observed by gender and on
each island.

8.4   QQ-Plots

Normal probability QQ-plots provide a means for comparing the distribution
of a sample against the standard normal distribution, the relevant function
along with arguments of interest being

```
qqnorm(y, main = "Normal Q-Q Plot",
     xlab = "Theoretical Quantiles",
     ylab = "Sample Quantiles", ...)
```

Here y represents the sample in question, which must be passed into the
function call. All other arguments listed have default values which can be
changed if so desired. Among other arguments that might find use and can be
included in a qqnorm function call are xlim, ylim, sub, and various applicable
graphics parameter setting arguments.

Alternatively, a sample's distributional properties can be compared against
any other theoretical distribution using

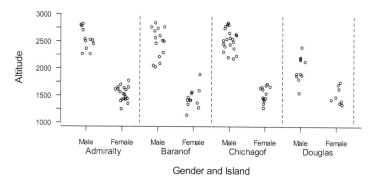

FIGURE 8.12
This figure is equivalent to Figure 8.11 in all respects except in that it requires a single `stripchart` function call, as opposed to four.

```
qqplot(x, y, xlab = deparse(substitute(x)),
    ylab = deparse(substitute(y)) ...)
```

Here y represents the sample quantiles and x represents the theoretical quantiles to be used for the comparison. Both x and y are required, and all other arguments listed are optional. Here too the axes limits can be defined through `xlim` and `ylim`, and the various labels and titles can be assigned through the use of `xlab`, `ylab`, `main`, and `sub`.

8.4.1 Normal Probability QQ-Plots

It is useful to know what is behind the construction of QQ-plots, particularly if the more generic `qqplot` function is to be used. The following discussion addresses the process involved in the construction of a *normal probability QQ-plot*. This can be extended to QQ-plots of other distributions through analogous steps.

Denote the sample of interest by $\{y_1, y_2, \ldots, y_n\}$, and begin by sorting this sample in ascending order. Denote the sorted data by $\{y_{(1)}, y_{(2)}, \ldots, y_{(n)}\}$, these are referred to as the *sample* or *observed quantiles*. Next, compute what are referred to as *plotting* (or *probability*) *points*, p_i. In R these plotting points can be computed using the `ppoints` function,

```
ppoints(n, a = ifelse(n <= 10, 3/8, 1/2))
```

The two default formulas used by `ppoints` are: For $i = 1, 2, \ldots, n$,

$$p_i = \begin{cases} (i - 3/8)/(n + 1/4) & \text{if } n \le 10, \\ (i - 1/2)/n & \text{if } n > 10. \end{cases}$$

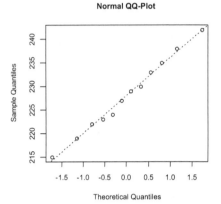

FIGURE 8.13
Normal probability QQ-plot of the sample `normal` from the list `qqData`. The y-intercept and slope of the reference line are the sample mean and standard deviation, respectively.

See, for example, [128, p. 33] for a discussion on other formulas used to compute plotting points specifically for use in constructing normal probability QQ-plots.

The plotting points, p_i, are then used to compute theoretical quantiles, x_i, corresponding to each sorted sample quantile, $y_{(i)}$. For normal probability QQ-plots this involves finding x_i for $i = 1, 2, \ldots, n$ such that $P(X \leq x_i) = p_i$, where $X \sim N(0, 1)$. The x_i are referred to as the *theoretical*, or *expected quantiles*. These quantiles are computed using the `qnorm` function with all arguments in the default setting (see Section 11.1.3) and then the ordered pairs $(x_i, y_{(i)})$ are plotted using the `plot` function. A reference line can be superimposed on the resulting figure. Preferred axes labels (`xlab` and `ylab`), main and subtitles (`main` and `sub`) can be included in the `plot` function call as desired.

The above process (see script for this section) when applied to the sample `normal` in the list `qqData` produces Figure 8.13. This figure is duplicated using the functions `qqnorm` and `abline` as follows.

```
> qqnorm(y = qqData$normal)
> abline(a = mean(qqData$normal),
+        b = sd(qqData$normal), lty = 3)
```

The line $y = \bar{y} + s_y x$ is used for reference purposes; see Section 8.4.3 for more on this, and Section 9.2 for more on the function `abline`.

8.4.2 Interpreting Normal Probability QQ-Plots

Figure 8.13 represents a somewhat idealized normal probability QQ-plot; all plotted points closely follow a linear trend. Such a normal probability QQ-plot suggests that the underlying random variable is very likely (at least approximately) normally distributed. Normal probability QQ-plots can provide additional information about data. Run the code for this section, applied to the samples in qqData, for the following discussions.

The Presence of Outliers

Outliers in a sample are those observations for which the plotted points lie in one or both of the tail ends of the data and away from the trend line formed by the bulk of the observations. The normal probability QQ-plot of the sample outs in qqData is an example of such cases. This behavior may or may not be a consequence of the underlying random variable not being normally distributed; valid but extreme values can occur in empirical distributions.[3]

Sometimes the removal of extreme values will improve the appearance of a QQ-plot. However, caution must be exercised when considering the removal of outliers from a dataset, and such measures should not be taken without valid justification. Moreover, though a QQ-plot does provide a means for flagging potential outliers it is advisable to use a combination of exploratory tools when investigating the presence of potential outliers.

Once identified, potential outliers should be re-examined from the source to eliminate the possibility of a data gathering or entry error. It is also suggested that outliers can sometimes be the most interesting of the observed data.

Gaps

Another feature appears in the form of gaps in the observed data; run the code for the QQ-plot of of the sample gaps in qqData. Gaps or jumps in the plotted points may be the result of rounding, or they may suggest the data are not representative of the range of possible data values. It is also possible that points on one side of such a gap might actually represent outliers, or that the sample actually represents a *mixed distribution.*

Step-like Patterns

Another pattern which may show up in a QQ-plot, but which is often easily correctable, involves the manner in which the data are recorded or computed; see the two QQ-plots of the steps sample in qqData. The left-hand plot produced by the code for this part is of the original data, and the plot on its

---

[3]If $X \sim N(0, 1)$, then $P(|X| \geq 2) \approx 0.0455$ and $P(|X| \geq 3) \approx 0.0027$. These suggest that when sampling a normally distributed population, there is a chance of approximately 1 in every 25 observations being at least 2 standard deviations away from the mean, and a chance of approximately 1 in every 500 observations being at least 3 standard deviations away from the mean.

right is of the same data rounded to two decimal places; a step-wise pattern appears when the data are rounded. While clusters do appear in the left hand plot, the step-wise structure is almost absent.

This suggests that if the difference among raw data values is small it may be appropriate to increase the decimal place representation of the data. It may also be possible rescale the data in a manner that increases the differences between observations that are close in value under the original scale.

Deviations from Symmetry

A normal distribution is symmetric about the mean, so deviations from symmetry imply a deviation from normality. In a normal probability QQ-plot deviations from symmetry are indicated by the presence of a noticeable concave trend in the plotted points.

If the points in the plot appear to follow a curve that is concave up then the underlying distribution of the observed data is likely to be skewed to the right, implying a heavy right tail. If the points in the plot appear to follow a curve that is concave down then the underlying distribution of the observed data is likely to be skewed to the left, implying a heavy left tail. The normal probability QQ-plots for the samples `right` (right-skewed) and `left` (left-skewed) in `qqData` illustrate such cases.

Violations of symmetry in a sample can be of concern. In such cases it is sometimes possible the data (hence, underlying random variable) may be transformed to reduce or possibly even correct asymmetry (see, for example, [65, pp. 138–144]).

Deviations from Normal Spread

Normal probability QQ-plots of approximately symmetric samples can also be used to diagnose deviations from normal spread. Suppose that the trend curve in a normal probability QQ-plot possesses a point of inflection, exhibiting a change of concavity. If the concavity changes from down to up, moving left to right, then the underlying distribution of the observed data is likely to have tails that are heavier than those of a normal distribution. Conversely, if the concavity changes from up to down, moving left to right, then the underlying distribution of the observed data is likely to have tails that are lighter than those of a normal distribution. The normal probability QQ-plots for the samples `heavy` (heavy-tailed) and `light` (light-tailed) illustrate such patterns.

The indication of tails that are considerably heavier than those of a normal distribution may be of concern since such data typically have outliers.

Mixed Normal Distributions

Another possibility is that observed data may belong to a mixed distribution, one containing observations from possibly the desired (normal) distribution as well as from one or more other distributions. These *contaminant distributions* may also be normal, but they may have means and/or variances that differ from each other.

Consider, for example, the random samples `norm1` and `norm2` in `qqData` which are from two normal distributions having different means and variances. The individual normal probability QQ-plots of these samples exhibit linear trends; see the top two QQ-plots in the figure produced by the code for this part, suggesting the underlying random variables are very likely normally distributed. However, the lower two plots indicate that if the two samples are combined and viewed as a single sample then the QQ normal probability plot does not exhibit a linear trend.

As these plots suggest, normal probability QQ-plots of normal mixtures can often exhibit patterns that might be mistaken for deviations from symmetry, spread, or other symptoms. See, for example, [128, Chapter 11] for more on mixed normal distributions.

8.4.3 More on Reference Lines for QQ-Plots

In normal probability QQ-plots, if the data are *not* standardized using the transformation $z = (y - \bar{y})/s$, a reference line of the form $y = \mu + \sigma z$ can be used where μ and σ represent the mean and standard deviation of the proposed theoretical normal distribution of y against which the distribution of the observed data is to be compared. The values of μ and σ are estimated using the sample mean and standard deviation of the observed data. To get such reference lines inserted in a normal probability QQ-plot, code of the form

```
abline(a = mean(y), b = sd(y), lty = 2)
```

can be used. Here y represents the test sample and `lty = 2` produces a dashed line.

If the data are standardized, then the line $y = x$ serves as a suitable reference line. This particular approach is useful if the observed data values have large absolute values. In this case

```
abline(a = 0, b = 1, lty = 2)
```

plots the line.

An alternative reference line can be superimposed on a current QQ-plot using the function

```
qqline(y, distribution = qnorm,
    probs = c(0.25, 0.75), qtype = 7, ...)
```

where y is assigned the sample used to construct the QQ-plot, and `distribution` is assigned a function (name) that is to be used to compute the relevant theoretical quantiles. The argument assignments `probs = c(0.25, 0.75)` and `qtype = 7` are passed into the `quantile` function to compute the

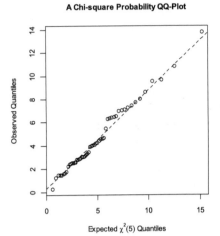

FIGURE 8.14
A chi-square probability QQ-plot with $\nu = 5$ degrees of freedom. The choice of how to construct the QQ-line here is made by identifying that 50%-portion of plotted points that appear in the highest concentration.

first and third quartiles which are then used to find the intercept and slope of the reference line to be used.[4] This (default) setting, to extract the first and third quartiles, is for normal probability QQ-plots; see the next section for cases in which other comparison distributions are used.

8.4.4   QQ-Plots for Other Distributions

As demonstrated in two of the examples in the documentation for the function qqplot, QQ-plots can be constructed for distributions other than the normal distribution. Run these examples and look at the second (uses a t-distribution) and last (uses a chi-square distribution) examples. Here are a few more details, using the the last example as an illustration.

First, create a sample that can be replicated by setting the random-number generator seed, then generate a vector of length 500 from the chi-square distribution with 5 degrees of freedom using the function rchisq.[5]

```
> set.seed(seed = 23); y <- rchisq(n = 500, df = 5)
```

---

[4]In the default setting the function qnorm with default parameter settings is assigned. See the next example in Section 8.4.4 where the comparison distribution is not the standard normal distribution.

[5]See Section 11.1.4 for details on R's chi-square distribution functions.

These represent the "observed" quantiles (a simulated sample). Next, obtain corresponding expected quantiles from the theoretical distribution with the help of the function `ppoints`,

```
> x <- qchisq(p = ppoints(n = 500), df = 5)
```

Now construct the QQ-plot—note the use of the `expression` function to insert *mathematical annotation* in `xlab`; see Section 10.3 for more on this.

```
> qqplot(x, y,
+     main = "A Chi-square Probability QQ-Plot",
+     xlab = expression("Expected"~{chi^2}(5)~"Quantiles"),
+     ylab = "Observed Quantiles")
```

Finally, to include a QQ-line for reference, first construct a function whose name is to be assigned to the argument `distribution` in the `qqline` function call.

```
> chisquare <- function(p){qchisq(p, df = 5)}
```

Then, in the above constructed chi-square probability QQ-plot, observe that the highest concentration of plotted points appear in the lower-left corner of the plot. So, one might consider assigning probabilities of 0 and 0.5 to the `probs` argument in the `qqline` function call.

```
> qqline(y, distribution = chisquare,
+     probs = c(0, 0.5), lty = 2)
```

The result appears in Figure 8.14. See Section 17.1 for an application of such QQ-plots to assessing bivariate normality, and a reason for the 50% region chosen. The script file contains sample code to construct an alternative reference line for such QQ-plots.

8.5   Additional Notes

In this section some further capabilities of the `ifelse` and `axis` functions are explored. Refer to the script file to follow the discussions given below.

8.5.1   More on the `ifelse` Function

The `ifelse` function was first seen in Section 4.2 in converting numeric data into categorical data, and in Section 8.2 to customize labels for the axis in Figure 8.6. The usage definition for this function is

```
ifelse(test, yes, no)
```

where the argument `test` is an object which is of (or can be converted into) logical mode. Then, the value (or evaluation) assigned to `yes` is returned for all elements of `test` that are TRUE, and the value (or evaluation) assigned to `no` is returned for all elements of `test` that are FALSE. The class of object `test` is either of logical mode or is one that can be coerced into logical mode. Here are some preliminary examples.

Creating a Two-Level Character Vector

In Section 4.2 this function was used in the manner

```
ifelse(test = (numbers < median(numbers)),
    yes = "lower", no = "upper")
```

where the numeric vector `numbers` was compared against the median of the contents of `numbers`. So the `test` object, `numbers < median(numbers)`, ends up being a logical vector. In this case the `ifelse` function returns the value `"lower"` for each TRUE in `test`, and `"upper"` for each FALSE.

Selective Tickmark Labels for an Axis

In Section 8.2 this function was used in the manner

```
ifelse(test = (ticks%%2 == 0), yes = ticks, no = "")
```

Here, the entries of the numeric vector `ticks` modulo 2 are compared against 0 (see Section 3.2 for the definition of the modulus operator). The resulting `test` object is once again a logical vector. Now, while the value returned for FALSE entries in `test` is always the "empty" character `""`, the values returned for TRUE entries are the corresponding entries of the numeric vector `ticks`.

8.5.2   Revisiting the `axis` Function

Figures 8.15 and 8.16 provide a couple of further illustrations of how the `axis` function (and some other features) may be used to alter or enhance the appearance of figures.

The usage definition for this function, with a small selection of arguments of interest for the following illustrations, is

```
axis(side, at = NULL, labels = TRUE, line = NA,
    pos = NA, ...)
```

The only required argument for this function is `side` (1 for the bottom axis; 2 for left-hand; 3 for top; and 4 for the right-hand axis). The argument `at` is an optional argument that can be assigned a vector of values that identify the

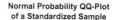

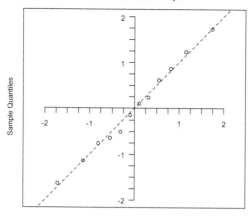

FIGURE 8.15

Example of a plot in which the axes are placed in the traditional locations through the inclusion of `pos = 0` in both of the `axis` function calls. Also, the aspect ratio is adjusted through the inclusion of `asp = 1` in the `qqnorm` function call, and margins are altered to maximize the plotting area through the graphics parameter `mar` in the `par` function call.

locations along the axis in question at which tickmarks are to be drawn. By default, these values are determined from the data used in the construction of the plot in question.

The argument `labels` is also an optional argument that can be assigned a vector containing the desired labels to be placed at each tickmark. By default, these labels are taken from the vector `at`, or the data used in the plot construction.

The argument `line` can be used to specify the margin line on which to place the axis, the default position is on `line = 0` for the side in question. Finally, the argument `pos` can be used to identify the coordinate on the orthogonal axis at which the axis line is to be drawn; see Figure 8.15 and corresponding code in this chapter's script file.

As previously demonstrated, the argument `las` can be used (as one of the graphics parameters "...") to set the orientation of the tickmark labels. Assignment options for `las` include: 0 for labels parallel to the axis in question; 1 for horizontally aligned tickmark labels; 2 for labels perpendicular to the axis in question; and 3 for labels that are oriented vertically.

Figure 8.16 includes features from the histograms constructed in Section 8.1. This is done through a left-hand axis representing the frequencies at which

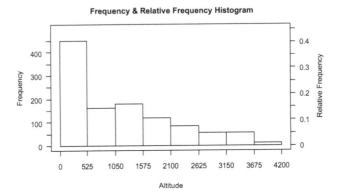

FIGURE 8.16

Taking advantage of scaled axis labels to use a single histogram to represent both a frequency and a relative frequency histogram.

deer were observed at various altitudes (using the vector **altitudes** extracted earlier in this chapter from the data frame **deerStudy**), as well as a right-hand axis representing the relative frequencies.

The right-hand axis is constructed and labeled by simply manipulating the vectors assigned to the arguments **at** and **labels** for the left-hand axis.

Here are portions of the code used for the right-hand axis (see script file for complete code). First create a histogram object, suppressing the plot, and extract the bin frequencies from the resulting object.

```
> histObj <- hist(x = altitude, plot = FALSE,
+     breaks = seq(from = 0, to = 4200, length.out = 9),
+     include.lowest = TRUE, right = FALSE)
> freqs <- histObj$counts
```

Now, focusing entirely on computations relevant to the right-hand axis, create "pretty" relative frequency tickmark locations.

```
> rfLocs <- pretty(x = freqs/sum(freqs), n = 10)
```

Using the numbers in **rfLocs**, create selective relative frequency tickmark labels.

```
> rfLabs <- ifelse(test = ((rfLocs*100)%%10 == 0),
+     yes = rfLocs, no = "")
```

Then, to plot the right-hand axis, the code used is

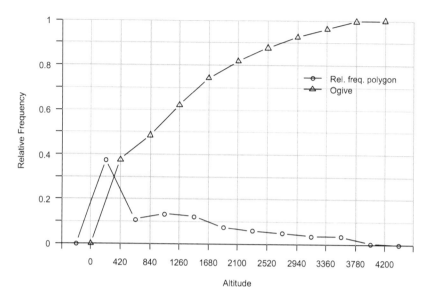

FIGURE 8.17
Relative frequency polygon and ogive of `altitude` data with gridlines.

```
> axis(side = 4,
+      at = rfLocs*sum(freqs), labels = rfLabs, las = 1)
```

Note the use of the `pretty` function in computing tickmark locations. This function uses numbers in the vector assigned to the argument x to compute an approximate desired number, n, of "nice" tickmark locations. This is a convenient and effective alternative to creating your own tickmark locations from scratch using, for example, the `seq` function or by observation.

8.5.3 Frequency Polygons and Ogives

Frequency polygons and ogives (see Figure 8.17) can be easily constructed through ideas introduced in the previous illustrations, and by taking advantage of the function `lines` (see Section 9.2).

For any given application of the `hist` function let $b_1, b_2, \ldots, b_n$ represent the equally spaced breaks, $m_1, m_2, \ldots, m_{n-1}$ the class midpoints, and $f_1, f_2, \ldots, f_{n-1}$ the class counts used by the function to construct the desired histogram. Next, with w being the class width, for $i = 0, 1, 2, \ldots, n$, define

$$x_i = \begin{cases} m_1 - w & i = 0 \\ m_i & 1 \leq i \leq n-1 \\ m_{n-1} + w & i = n \end{cases} \quad \text{and} \quad y_i = \begin{cases} 0 & i = 0 \\ f_i & 1 \leq i \leq n-1 \\ 0 & i = n \end{cases}$$

Then a frequency polygon is obtained by plotting lines between the ordered pairs (x_i, y_i), $i = 0, 1, \ldots, n$. Equivalently, a relative frequency polygon is obtained by using relative frequencies, say rf_i, in place of frequencies, f_i, in the definition of y_i.

Next, for $j = 1, 2, \ldots, n$, define

$$x_j = b_j \quad \text{and} \quad y_j = \begin{cases} 0 & j = 1 \\ cf_j = \sum_{i=1}^{j} f_i & j \neq 1 \end{cases}$$

An ogive is obtained by plotting lines between the ordered pairs (x_j, y_j), $j = 1, 2, \ldots, n$, and an ogive using cumulative relative frequencies is obtained by using relative cumulative frequencies, say crf_j, in place of frequencies, cf_j, in the definition of the y_j.

Figure 8.17 provides an example of using the `altitude` data to plot a relative frequency polygon and a cumulative relative frequency ogive on the same axes. The histogram object used to provide classes and frequencies for this plot differs slightly from that used in the previous example, and the gridlines are plotted using the `segments` function described in Section 10.4.2. See the script file for code used to construct this figure.

9

Scatterplots, Lines, and Curves

The main focus of this chapter is on plots associated with bivariate numeric data such as scatterplots, lines, and curves. Further examples of how the appearances of figures can be customized are included, along with using ideas covered to construct time-series plots. The script file for this chapter, `Chapter9Script.R`, contains quite a few additional examples and exploratory code; use it to follow along with the discussions and ideas presented.

9.1 Scatterplots

Scatterplots of bivariate numeric data having a *response (dependent) variable*, say Y, and an *explanatory (independent) variable*, say X, involve plotting ordered pairs (x_i, y_i), $i = 1, 2, \ldots, n$, as points on the 2-dimensional coordinate axes. The typical use of such plots is in either exploring the manner in which the observed responses vary in relation to variations in the explanatory variable, or providing a graphical display of a known relationship between two variables.

The function that finds use in this area is the `plot` function, the usage definition of which, along with arguments of interest, is

```
plot(x, y = NULL, type = "p", xlim = NULL, ylim = NULL,
     main = NULL, sub = NULL, xlab = NULL, ylab = NULL,
     axes = TRUE, frame.plot = axes, asp = NA, ...)
```

The arguments x and y represent variables for the horizontal and vertical axes, respectively. Among the remaining arguments listed are the familiar axes limits, labeling and title arguments used in previous graphics functions. The aspect ratio argument `asp` was seen used for Figure 8.15 in the previous chapter.[1] The arguments `axes` and `frame.plot` have also been seen in

---

[1]See the script file for a demonstration of the effect of altering the aspect ratio. Very briefly, suppose `asp = a/b` and suppose 1 inch along the horizontal axis represents a units in the horizontal scale, then 1 inch along the vertical axis represents b units in the vertical scale.

constructing earlier plots, and the argument `type` specifies the type of plot constructed. Of interest for this section is `type = "p"`, the default.[2]

An alternative approach to entering the data to be plotted is to pass them into the function in `formula` format,

```
plot(formula, data, subset, ...)
```

The two additional arguments that appear here, `data` and `subset`, are used as previously seen.

9.1.1 Basic Plots

The data used for the first few examples below are contained in the data frame `stopDist`. In this data frame the vector `distance` contains the stopping distance for a car travelling at the corresponding speed in the vector `speed`. The entries of the factor `driver` identify one of two drivers, and the entries of `make` identify the car (one of two) as being an old or new model. Before beginning the following illustrations find the range of values in the vectors `distance` and `speed`; knowing this is helpful in deciding what to use for axes limits if a custom scale is desired.

Constructing a Scatterplot

Using the formula approach, a scatterplot of stopping distance can be constructed, see Figure 9.1(a), by running code of the form

```
> plot(formula = distance ~ speed, data = stopDist,
+       xlim = c(5, 25), ylim = c(0, 120),
+       xlab = "Speed (mph)", ylab = "Distance (ft)",
+       main = "Stopping Distance vs. Speed",
+       sub = "(a) Default tickmarks")
```

This figure can be duplicated using the `x, y` argument assignment approach.

Customizing the Axes

If so desired, the axes in this scatterplot can be customized as in Figure 9.1(b). To accomplish this, first duplicate Figure 9.1(a), but with a different subtitle. Next, prepare a vector containing locations of additional tickmarks for the horizontal axis, and then insert the additional tickmarks with no labels.

```
> xlocs <- seq(from = 5, to = 25, by = 1)
> axis(side = 1, at = xlocs, tcl = -0.2,
+       labels = rep(x = "", times = 21))
```

---

[2]Other options include lines (`type = "l"`), and both lines and points (`type = "b"`).

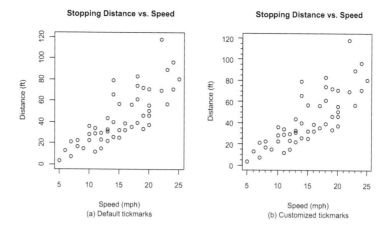

FIGURE 9.1
Scatterplots of stopping distance against speed, one with default tickmarks and tickmark labeling, and the other with customized tickmarks and tickmark labeling.

Note that in the previously drawn axis, tickmarks and tickmark labels are unaffected, and the graphics parameter setting `tcl = -0.2` reduces the length of the newly inserted tickmarks (the default setting is `tcl = -0.5`). The inserting of further tickmark labels is prevented by assigning the empty character, `""`, to the tickmark labels. Similarly, enhancements can be added to the vertical axis (see the script file).

Identifying Plotted Points

Let the arguments x and y represent the *x*- and *y*-variables for the plot in question, then the function

```
identify(x, y = NULL)
```

can be used as follows. First run code to duplicate Figure 9.1(b), then run

```
> with(data = stopDist,
+     expr = identify(x = speed, y = distance))
```

Now, move the cursor onto the graphics device window, cross-hairs will appear. Move the cross-hairs over to the points of interest, in this case the two upper points that lie outside of the bulk of the plotted points and left-click the mouse. When this is done the case number for each plotted ordered pair is inserted into the image right next to the plotted point of interest. To exit the process, right-click the mouse and select "Stop"; see Figure 9.2.

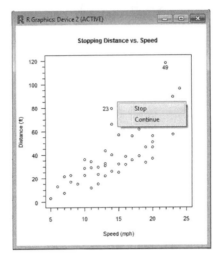

FIGURE 9.2

Identifying plotted points of interest by their case numbers using the `identify` function.

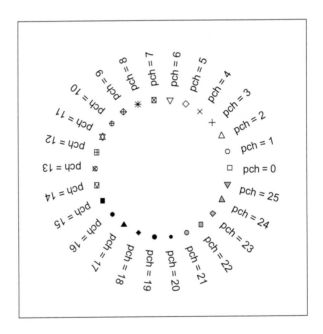

FIGURE 9.3

The complete list of special plotting characters (`pch`) that can be used to define symbols used to plot points. The fillable symbols 21–25 have been shaded gray. See script file for code used to construct this figure.

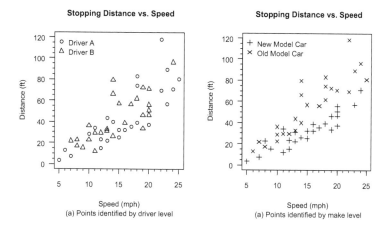

FIGURE 9.4

Using different symbols for plotting characters to distinguish factor levels to which points belong.

9.1.2 Manipulating Plotting Characters

Any single character can be used to plot a point, and other symbols can be plotted through the assignment of an integer from 0 through 25 to the `pch` graphics parameter; see Figure 9.3. Only the symbols 21–25 can be filled with any background color of choice.

Plotting character symbols are useful in scatterplots if it is wished to identify the plotted points by factor levels. See, for example the two plots in Figure 9.4 and the code in the script file.

9.1.3 Plotting Transformed Data

There are occasions when a scatterplot of transformed data is desired. In such cases the transformed data can be passed into the `plot` function by means of two new vectors created *outside* the `plot` function call, or the transformation can be performed *inside* the `plot` function call.

Transforming Both x and y Using Built-in Functions

For Figure 9.5 both the vertical and horizontal scales are transformed using the natural logarithm and the `plot` function call used is

```
> with(data = stopDist,
+     expr = plot(x = log(speed), y = log(distance),
+         xlim = c(1.5, 3.5), ylim = c(1, 5),
+         xlab = "ln(Speed)", ylab = "ln(Distance)",
```

```
+           main = "Transformed Stopping Distance Data",
+           las = 1))
```

Then, minor tickmarks are included as in previous scatterplots. The formula approach can also be used. In place of the argument assignments to x and y, use the argument assignment `formula = log(distance) ~ log(speed)`.

Transforming x Using a Computational Formula

Suppose a plot of the breaking distance against the square of the speed is desired. Consider using the `formula` method,

```
> plot(formula = distance ~ I(speed^2), data = stopDist,
+      xlab = expression((Speed)^2), ylab = "Distance")
```

Notice how the transformed data for `speed` are passed into the function; the *inhibit* function, `I`, has to be used here.

Sample exploratory code in the script for this section show that the use of the inhibit function when transforming the independent variable is not always necessary in the `plot` function for the formula method, and is not needed at all for the default method (through the arguments x and y). In the case of transforming the dependent variable, the inhibit function is not needed for either plotting method.

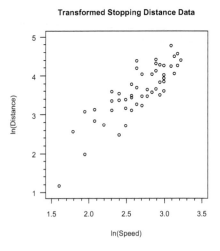

Transformed Stopping Distance Data

FIGURE 9.5
Scatter plot of stopping distance data where both variables are log-transformed.

Pairwise Plots of Weight, Altitude and Inland by Gender

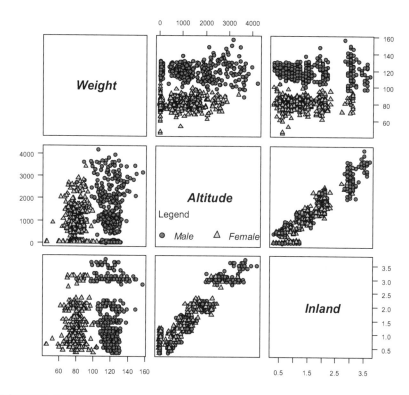

FIGURE 9.6

A matrix scatterplot of the variables `weight`, `altitude` and `inland` from the
data frame `deerStudy`. Different plotting characters are used for each gender.

9.1.4 Matrix Scatterplots

Matrix scatterplots are useful for simultaneously viewing scatterplots of variable pairs appearing in multivariate datasets, such as in 9.6. Two functions can
be taken advantage of, the versatile `plot` function, or the specialized `pairs`
function.

If the data appear in a data frame whose columns are all numeric, for example `weight`, `altitude` and `inland` in the data frame `deerStudy`, extracted
using

```
> theData <- subset(x = deerStudy,
+     select = c("gender", "weight", "altitude", "inland"))
```

then a default matrix scatterplot can be obtained by running

```
> plot(x = theData[, -1])
```

or using `pairs` in place of `plot`. The purpose of the "-1" is to exclude `gender` from the data passed into the function.

Figure 9.6 is equivalent to the figure produced by the previous `plot` and `pairs` function calls except in that it provides a bit more detail with plotted points being identified by gender. See the script file for code with line-by-line details.

More on the `pairs` Function

The `pairs` function has arguments that provide some additional capabilities. The usage definition for this function, along with some arguments of interest, is

```
pairs(x, labels, panel = points, ...,
    lower.panel = panel, upper.panel = panel,
    diag.panel = NULL, text.panel = textPanel,
    label.pos = 0.5 + has.diag/3, line.main = 3)
```

As the previous `pairs` function calls suggest, x is the only required argument and the `labels` are obtained from the data frame (or matrix) column names. As the R documentation example code for this function suggest, quite a bit of flexibility exists for the `pairs` function through functions assigned to the arguments `panel`, `lower.panel`, `upper.panel` and `diag.panel`. Exploratory code for these are given in the script for this section.

9.1.5 The `matplot` Function

This function can be used to superimpose multiple scatterplots on the same axes. The usage definition along with arguments of interest is

```
matplot(x, y, type = "p", pch = NULL, col = 1:6, bg = NA,
    xlab = NULL, ylab = NULL, xlim = NULL, ylim = NULL,
    ..., add = FALSE)
```

Note that the argument `bg` is used to assign background colors *only* to the open plotting characters `pch = 21:25`, and is assigned a vector of background (fill) colors for the symbols plotted. The default `NA` results in the plotted symbols being open (that is, without any colors filled in). Also useful to note is that the plot `type` can be points (`type = "p"`), or lines (`type = "l"`), or both lines and points (`type = "b"`). All other arguments listed are as for previously seen graphics functions. If `add = FALSE`, the `matplot` function opens a new

plotting window, or replaces the contents of an existing window. If `add =`
`TRUE` then the plotted points are added to an existing active plot.

There are two ways in which the data to be plotted may be passed into
the function through the arguments x and y.

When x Is a Vector and y Is a Matrix

One way is to use x to define a single vector containing the (common) x-
values. Then y is used to introduce the corresponding observed response values
in the form of a matrix having two or more columns and the same number of
rows as the length of x. The columns of y can be thought of as representing a
collection of samples of observed responses belonging to different experimental
units. So, if for each $i = 1, 2, \ldots, n$ and $j = 1, 2, \ldots, p$, y_{ij} represents the
observed response in the jth experimental unit corresponding to x_i, then the
ordered pairs that are plotted are (x_i, y_{ij}).

Consider the data frame `grass` which contains data on the growth of grass
under five treatments (four types of grass fertilizer and one with no fertilizer,
the control) observed over a period of 14 days. To get the experiment started,
grass is grown in five lots prepared with identical topsoil. Once the grass in all
lots exceed 5 cm in height, all lots are mowed to a height of 5 cm at the end of
that day (day zero). The first column, `day`, contains the day (starting at day
zero) on which the length of grass under each fertilizer treatment is observed.
The remaining columns contain averages of the observed heights under each
treatment at the end of each day. Figure 9.7 (a) is the result of the following
code.

First, for convenience, create variable identifying vectors,

```
> indepVar <-"day"
> depVars <- c("control", "A", "B", "C", "D")
```

Then, after preparing a suitable plotting window, run

```
> matplot(x = grass[, indepVar], y = grass[, depVars],
+     pch = 0:4, col = "black", ylim = c(0, 20),
+     xlab = "Day", ylab ="Grass Height (cm)", sub = "(a)")
> legend(x = "topleft", legend = depVars, pch = 0:4)
```

Code for variations of this plot (using the other two plot types) are contained
in the script for this section.

When Both x and y Are Matrices

Suppose the ordered pairs for the data to be plotted have the form
(x_{ij}, y_{ij}), where $j = 1, 2, \ldots, p$ identifies the treatment and $i = 1, 2, \ldots, n$
identifies the observation (case) number. In this case both x and y are $n \times p$

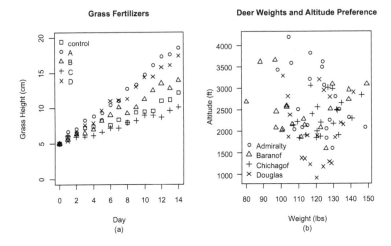

FIGURE 9.7
Superimposing multiple scatterplots on the same axes using the `matplot` function.

matrices, the jth column of x being paired with the jth column of y. Such data might be expected from observational studies.

For example, consider performing a graphical exploration of whether the heavier a male Sitka Blacktail deer on the "ABCD" islands, the higher up it tends to hang out in the months of June through August. Data for random samples of 20 male deer from each island, extracted from the data frame `deerStudy`, are stored in the list `howHigh`.

Since (in theory) such data would be obtained from an observational study, the samples of observed weights for the four islands are (pretty much) guaranteed to differ. So, both of the arguments x and y are assigned matrices.

Figure 9.7(b) is constructed by first obtaining the scatterplot using

```
> with(data = howHigh,
+      expr = matplot(x = weights, y = heights,
+          pch = 1:4, col = "black", type = "p", sub = "(b)",
+          xlab = "Weight (lbs)", ylab ="Altitude (ft)",
+          main = "Deer Weights and Altitude Preferences"))
```

To wrap things up, include the legend for the plotting characters used to represent each island.

Notice that while Figure 9.7(a) is useful in that clear and apparently distinct trends across treatments are evident, easily discernable trends are not present in Figure 9.7(b).

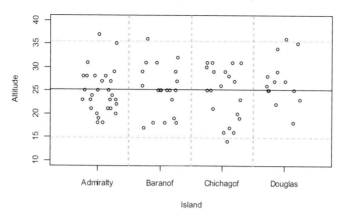

FIGURE 9.8

Plotting horizontal and vertical lines for purposes of reference. Horizontal lines may be used to assess homogeneity in spread and/or flag potential extreme values, and the vertical lines here serve as separators between factor levels.

9.2 Graphs of Lines

The function for this purpose is `abline` which can be used to include horizontal, vertical, and sloped lines in a plot. The usage definition for this function with selected arguments is

```
abline(a = NULL, b = NULL, h = NULL, v = NULL, ...)
```

A vector of values passed to the argument `h` causes horizontal lines to be plotted, the argument `v` is used similarly for vertical lines.

For slanted lines, if the arguments `a` and `b` are each assigned a (single) value, a line with intercept `a` and slope `b` is plotted.

Line types (`lty`), line colors (`col`), and line width (`lwd`) can be specified as desired, either through a vector of types/colors/widths or as single values.

Horizontal and Vertical Lines

Figure 9.8 provides one example of how horizontal and vertical lines might find use in a plot. The construction of this figure first involved defining an appropriate plotting window, then constructing a stripchart of altitudes at which deer were observed on each island in January (see the script file for code). Next, calculate the overall mean (`xBar`) and standard deviation (`sDev`) of January altitudes and use altitude values that are two standard deviations away from the mean altitude for the reference lines in a vector.

```
> refLines <- c(xBar-2*sDev, xBar, xBar + 2*sDev)
```

Finally, insert the horizontal reference lines

```
> abline(h = refLines, col = c("gray", "black", "gray"),
+     lty = c(2, 1, 2))
```

and then insert vertical lines to partition plotted points by islands

```
> abline(v = c(1.5, 2.5, 3.5), lty = 3, col = "gray",
+     lwd = 2)
```

See the script file for an example of how the `abline` function can also be used to insert a grid for the rectangular plane.

Slant Lines

A previous application of the `abline` function appeared in the construction of the normal probability QQ-plot shown in Figure 8.13. To do the same for the vector `distance` in the data frame `stopDist`, run (figure not shown)

```
> with(data = stopDist,
+     expr = {qqnorm(y = distance)
+             abline(a = mean(distance),
+                    b = sd(distance), lty = 2)})
```

The arguments `a` and `b` cannot be assigned more than one value each.

9.3   Graphs of Curves

The use of two functions for constructing curves in the rectangular plane are demonstrated here, `curve` and `lines`.

The `curve` Function

The `curve` function uses a given function to compute ordered pairs and then plot line segments from one ordered pair to the next, either on an existing plot or in a new plotting window of its own. The usage definition of this function along with a selection of arguments of interest is

```
curve(expr, from = NULL, to = NULL, n = 101, add = FALSE,
    xname = "x", xlab = xname, ylab = NULL, ...)
```

The argument `expr` is assigned either the name of a function, or an expression written in function form, or a function call of x (or whatever variable name is assigned to the argument `xname`). The arguments `from` and `to` provide the horizontal range over which the curve is to be plotted, and `n` is an integer specifying the number of ordered pairs that are to be used in constructing the plot (the default setting of `n = 101` typically provides for adequately smooth looking curves). The axes labels `xlab` and `ylab` can be left in their default settings, or not, as in applications of previous plotting functions.

If the curve is to be superimposed on (or added to) an existing plot, use `add = TRUE`. Otherwise, the curve is plotted in its own plotting window. The graphical parameters `lty`, `lwd` and `col` can be used as before.

The following examples, see Figure 9.9, demonstrate two ways in which the `curve` function can be taken advantage of.

Example: Superimposing a Curve on an Existing Plot

For the first example, see Figure 9.9 (a), the `curve` function is used to plot a curve that represents a model to which the `distance` and `speed` data from `stopDist` are fitted.

The scatterplot is constructed, then the data are fitted to a desired model (see script for this case and, for example, [65, pp. 134–138]) so as to be able to create the function that describes the fitted model,

```
> f <- function(x) {0.6252*x^1.5288}
```

Now use the `curve` function to add this curve to the plot and finally include a legend.[3]

```
> curve(expr = f(x), from = 0, to = 30, add = TRUE,
+     lty = 2)
> legend(x = "topleft", lty = 2, bty = "n",
+     legend = expression(y == 0.6252*x^1.5288))
```

Example: Plotting a Curve in Its Own Window

The second example, see Figure 9.9 (b), is simply the density curve for the F-distribution with $\nu_1 = 5$ and $\nu_2 = 35$ degrees of freedom in its own window. This is constructed using

---

[3]See Section 10.3 for more on plotting mathematical annotation.

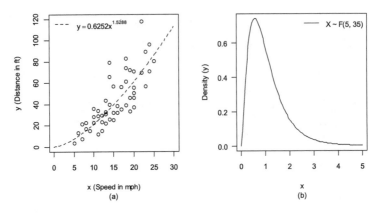

FIGURE 9.9

Two basic applications of the `curve` function: (a) Inserting a curve in an existing plot; (b) Plotting a curve in its own window.

```
> curve(expr = df(x, df1 = 5, df2 = 35),
+     from = 0, to = 5, add = FALSE, las = 1,
+     xlab = "x", ylab = "Density (y)", sub = "(b)")
> legend(x = "topright", lty = 1, bty = "n",
+     legend = "X ~ F(5, 35)")
```

If so desired, density curves for multiple F-distributions can be superimposed on the same plot; see Section 9.4.

The `lines` Function

The `lines` function takes in ordered pairs and then plots line segments from one ordered pair to the next in an existing (active) plot. The usage definition of this function is

```
lines(x, y = NULL, type = "l", ...)
```

where the ordered pairs can be passed into the function either as an $n \times 2$ matrix through the argument x (in which case y = NULL, the first column of x represents the x-variable, and the second column the y-variable), or as two separate vectors of equal length through the arguments x and y. By default this function plots lines (through the argument assignment `type = "l"`), and the previously described graphical parameters `lty`, `lwd` and `col` can be used to serve the same purposes as described for previous plots.

While the plots in Figure 9.9 can be duplicated using the `lines` function (see code for this section in the script file), the main strength of the `lines`

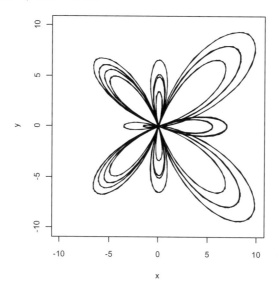

FIGURE 9.10
As opposed to the `curve` function, the `lines` function can be used to plot curves that do not explicitly describe y as a function of x.

function is the capability to construct fairly complicated curves for which a "nice" explicitly defined function (for y in terms of x) is not available.

Example: Plotting a Curve Using the `lines` Function

The example curve used here is not useful for any statistical application, but it does illustrate how a complex curve might be plotted. The curve in Figure 9.10 is defined in polar form by

$$r(t) = 2\,e^{\cos(t)} - 7\cos(4t) + 2 * \sin^3(t/4).$$

So for a give value of t, the corresponding ordered pair is given by

$$x = r(t)\cos(t), \quad y = r(t)\sin(t),$$

and for the illustration in question, $-10\pi \leq t \leq 10\pi$. Here is an outline of how Figure 9.10 is constructed (see the script file for code). The first task is to find appropriate vectors x and y. To do this define the polar function for r, then prepare a vector containing a large number of values for t (the more the merrier). Next, compute ordered pairs (x_i, y_i), activate the plotting window with a box around it, and set axes limits and aspect ratio. Next, include the axes with labels and, finally, plot the curve.

Which Function Should Be Used?

Based on the previous examples, the function choices at this point are the curve and lines. Here are some basic guidelines on function choices.

If all that is available is a set of (say n) ordered pairs, the lines function[4] is the obvious choice out of the two. If an actual equation defining the curve of interest is available, there are some considerations that should be kept in mind.

Broadly speaking, most (nice) curves in 2-space can be defined by an equation of the form $F(x, y) = 0$, and in many cases such equations can be solved for y so that y is expressed *explicitly* as a function of x, that is, $y = f(x)$. If this is not possible then y is said to be defined *implicitly* by the equation $F(x, y) = 0$.

The curve function is the more efficient method of plotting curves for cases where y can be expressed as a function of x, and as seen in the previous cases an argument assignment of the form expr = f(x) takes care of business.

For cases in which a curve cannot be defined explicitly, but which is defined implicitly by an equation of the form $F(x, y) = 0$, it is often possible to express the curve in parametric form. In such cases x can be expressed as a function of a third variable, say $x = x(t)$, and similarly $y = y(t)$. Varying the variable t from a starting value, say t_1, to an ending value, say t_n, then provides the set of ordered pairs (x_i, y_i), $i = 1, 2, \ldots, n$. Here the obvious function to use is the lines function (or the plot function with type = "l").

9.4 Superimposing Multiple Lines and/or Curves

Plotting multiple horizontal and/or vertical lines is best accomplished through appropriate calls of the abline function; see code for Figure 9.8. Similarly, slant lines whose slopes and intercepts are known can be plotted on the same axes using appropriate calls of the abline function.

For purposes of superimposing curves on the same axes, while multiple calls of the lines or curve function can be used to accomplish such tasks, the matlines function (a variation of the matplot function) serves the same purpose with a single function call.

Using the matlines, matplot or lines Functions

The usage definition for the matlines function is

```
matlines (x, y, type = "l", lty = 1:5, lwd = 1,
    col = 1:6, ...)
```

---

[4]Recall that the plot function with the argument assignment type = "l" also serves the same purpose for a *single curve plot*.

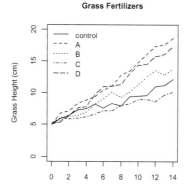

FIGURE 9.11
Superimposing multiple line graphs on the same axes using the `matlines` function.

The first two arguments for this function serve the same purposes as for the `matplot` function, and aside from the argument assignment `type = "l"` and the need for line related graphical parameter settings (of `lty` and `lwd`) the application is analogous to that of the `matplot` function.[5]

Consider constructing the "lines" equivalent of Figure 9.7 (a) using the `matlines` function; see Figure 9.11.

Unlike the `matplot` function and like the `lines` function, the `matlines` function requires an existing/current plotting window to be open before it can be used. So, a first step would be to activate and prepare the plotting window with a box around it (if so desired). Next, include the axes and axes labels; default settings for the axes tickmarks work fine here. And finally, the lines are plotted and the legend is included using

```
> matlines(x = grass[, 1], y = grass[, 2:6],
+     col = "black", lty = 1:5)
> legend(x = "topleft", lty = 1:5, bty = "n",
+     legend = c("control", "A", "B", "C", "D"))
```

Figure 9.11 can be duplicated using the `matplot` and the `lines` functions; see the script file for sample code.

---

[5] In fact, the `matplot` function can be made to duplicate a plot produced by the `matlines` function if `type = "l"` is used in place of `type = "p"`, and appropriate `lty`, `lwd` and `col` parameter settings are included in the `matplot` function call. See exploratory code in the script file for this section and the section on the `matplot` function for illustrations.

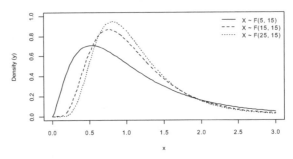

FIGURE 9.12
Superimposing multiple `curves` using the curve function, with the help of a for-loop. Note how the `paste` function is taken advantage of for the legend.

If both x and y are matrices, as was the case for Figure 9.7 (b), the process is analogous to the application with the `matlines` function. See the script file for constructing the "lines" equivalent of Figure 9.7 (b) using the `matlines` function.

Using the `curve` Function

Consider extending a figure such as Figure 9.9 (b) to enable a demonstration of the effect of varying ν_1, the degrees of freedom of the numerator of an F-distribution. With the help of a for-loop, the `curve` function can be taken advantage of as follows.

First prepare the plotting window, set the axes limits, plot the axes and include axes labels. Next, for convenience and brevity, prepare a vector, `dfN`, containing desired degrees of freedom for the numerator. Now plot the curves by looping through the desired degrees of freedom for the numerator.

```
> for (i in 1:3)
+ {   # Plot the density curve
+     curve(expr = df(x, df1 = dfN[i], df2 = 15),
+           from = 0, to = 3, lty = i, add = TRUE)
+ }   # End for-loop
```

Finally, insert a legend and the result appears in Figure 9.12.

Which Function Should Be Used?

Choosing which function to use for superimposing multiple curves on the same axes, as in previous such discussions, depends on the application in question as well as personal preference. Clearly, if sets of ordered pairs are

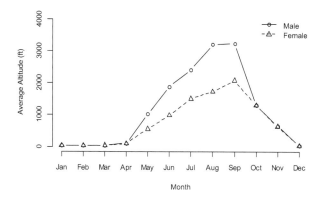

FIGURE 9.13
Time-series plots of average altitudes at which male and female deer are observed by month.

available, the simplest approach is to use the `matlines` (or `matplot`) function. Alternatively, if explicitly defined functions are available for the curves to be plotted, then the `curves` function is the more efficient way to go.

For equations in which the dependent variable cannot be written as a function of the independent variable, but which can be rewritten in parametric form, the approach described for Figure 9.10 can be employed using the `matlines` function (or the `matplot` or `lines` functions).

9.5   Time-Series Plots

Time-series data involve observed or measured values of a variable of interest over equally spaced time-intervals. It is fairly straightforward to work with time-series data using objects and tools already seen.

Consider looking at average altitudes, by gender, at which deer are seen each month of each year; see Figure 9.13. First place the needed information in a data frame.

```
> aveAltsGender <- with(data = deerStudy,
+     expr = tapply(X = altitude,
+         INDEX = list(month, gender), FUN = mean))
```

Get a suitable maximum and vertical axis scale for a plot of these altitudes using

```
> yMax <- ceiling(1.1*max(aveAltsGender)/500)*500
> yTicks <- seq(from = 0, to = yMax, by = 1000)
```

Next, prepare a window and construct the plot, with axes suppressed.

```
> matplot(x = 1:12, y = aveAltsGender, type = "b",
+     xlab = "Month", ylab = "Average Altitude (ft)",
+     pch = c(1,2), lty = c(1, 2), col = "black",
+     ylim = c(0, yMax), axes = FALSE)
```

Finally, insert the horizontal and vertical axes, slightly customized, as well as a legend in the usual manner.

To plot a single time-series, the only change needed in the above code would be to replace the `matplot` function by the `plot` function. Alternatively, a for-loop can be used (see script file for code) to construct multiple time-series plots such as in Figure 9.14.

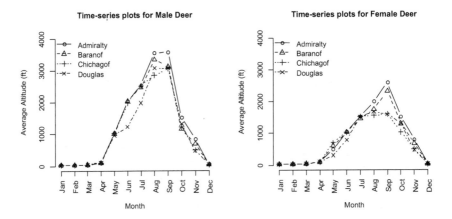

FIGURE 9.14

Time-series plots of average altitudes at which male and female deer were observed each month by islands.

10

More Graphics Tools

Some additional R functions that can be used to create a customized figure, or jazz up an existing figure are demonstrated, including a number of low level graphics functions that may be used to enhance a figure. The chapter closes with pulling together graphics ideas for an error bar function. Code used for this chapter, along with further illustrations, are provided in the script file for this chapter, `Chapter10Script.R`.

10.1   Partitioning Graphics Windows

The previously mentioned graphics parameters `mfrow` and `mfcol` work well when multiple figures with the same dimensions are to be placed in a single (partitioned) graphics window. However, there are occasions when different dimensions for the compartments in a partitioned window are more appropriate. Two functions that can be used for this purpose are the `layout` and the `split.screen` functions.

10.1.1   The `layout` Function

The relevant functions for this section, with arguments of interest, are

```
layout(mat, widths, heights)
layout.show(n = 1)
```

The `layout` function is used to define a plotting window configuration, and `layout.show` is used to view the configuration. The number assigned to the argument `n` in `layout.show` specifies the number of compartments to display, starting from the first.

An Example

A demonstration of using the `layout` function appears in the construction of Figure 10.1, which contains plots constructed from data in the previously encountered `stopDist` data frame. Here is one way in which this figure may be constructed.

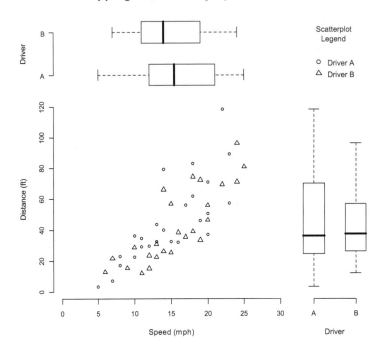

FIGURE 10.1

A demonstration of using the `layout` function to construct three different plots in the same plotting window.

First open a suitably sized graphics window (6 × 6 inches in this case) using any one of the graphics device functions described in Section 7.6.1. Then, using the `layout` function, partition this window into two columns with widths of 4.5 and 1.5 inches through the `widths` argument, and three rows with heights 0.5, 1.5 and 4 inches through the `heights` argument.

```
> layout(widths = c(4.5, 1.5), heights = c(0.5, 1.5, 4),
+     mat = matrix(data = c(5, 5, 1, 4, 2, 3), nrow = 3,
+         ncol = 2, byrow = TRUE))
```

To view the layout for the five screens and the plotting order specified through the `mat` argument, run

```
> layout.show(n = 5)
```

The matrix assigned to the `mat` argument places the first plot that is constructed in the left-hand plotting screen of the second row, the second in the bottom-left screen, and the third in the bottom-right screen. The right-hand screen of the second row is intended for the scatterplot legend, and the entire first row is reserved for the figure title. See the script file for the code used to construct this figure.

10.1.2   The `split.screen` Function

Here is a walk through an outline of the basic purpose of the `split.screen` function and its associated functions. The general usage definition is

```
split.screen(figs, screen, erase = TRUE)
```

where `figs` is the partition definition, `screen` identifies the plotting screen to be partitioned, and `erase` indicates whether the contents of the identified screen should be erased (or not) before partitioning the screen. There are three associated functions: `screen` (used to activate an identified screen), `erase.screen` (used to erase the contents of an identified screen), and `close.screen` (used to close one or more identified screens). There are two ways in which the partioning of a chosen screen can be accomplished.

Specifying the Number of Rows and Columns

After opening a suitably sized graphics window, the window can be split into x rows and y columns, in much the same way as was done with `par(mfrow = c(x,y))`, using a function call of the form

```
split.screen(figs = c(x, y), erase = TRUE)
```

The individual screens themselves (for example, screen k) can also be split into sub-screens using

```
split.screen(figs = c(a ,b), screen = k, erase = FALSE)
```

where k is the screen identifier starting with the first row, and moving across columns. To activate a particular screen for purposes of plotting, a function call of the form

```
screen(n = k, new = TRUE)
```

is used. The argument `new` is set to `FALSE` if a screen is opened later and further work is to be done on an existing plot. A chosen screen can be erased using

```
erase.screen(n = k)
```

and chosen screens, or all screens can be closed using versions of

```
close.screen(n, all = FALSE)
```

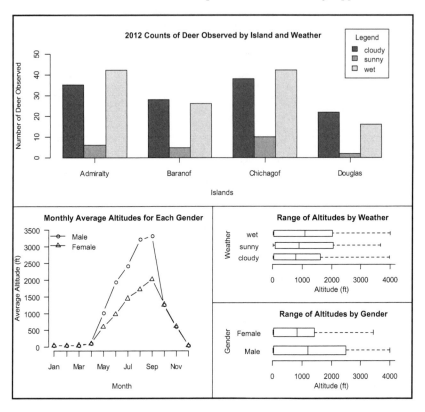

FIGURE 10.2
An illustration of using the `split.screen` function and the `figs = c(x, y)` feature to partition a plotting window. The framed outlines for each plot, and the whole figure are obtained using the `box` function.

An Example

After opening a suitably sized window, the partitioning for Figure 10.2 is accomplished as follows. First, the full screen is partitioned into two rows using

```
> split.screen(figs = c(2, 1))
```

and it will be observed that the resulting screens are numbered 1 and 2. Next, the lower screen (screen 2) is partitioned into two columns using

```
> split.screen(figs = c(1, 2), screen = 2)
```

resulting in screens that are numbered 3 and 4. Finally, the lower-right screen (screen 4) is partitioned into two rows using

```
> split.screen(figs = c(2, 1), screen = 4)
```

These are numbered 5 and 6. Knowing the numbers assigned to each subscreen is helpful in the plotting phase.

To procede with the plotting, each screen is first activated using the `screen` function. For example, to begin plotting the frequency bar chart in the top plotting screen `screen(n = 1)` is run first. Then the various plotting code is run.

A new plotting feature used here is the `box` function which finds use in plotting frames about plots and figures; see Section 10.4.3 for more details and see the script file for code used to construct Figure 10.2.

So, the `split.screen` function does provide quite a bit more power (than the `mfrow` and `mfcol` parameter settings) in designing the layout of plots in a graphics window.

Specifying the Side Positions of Each Plotting Screen

More control over the partition sizes can be achieved if the `figs` argument is assigned a matrix containing information about side positions in the horizontal and vertical.

Let `leftk`, `rightk`, `bottomk`, `topk` (for $k = 1, 2, \ldots, n$) denote the positions of the left, right, bottom and top sides, respectively, of figure k as proportions of the corresponding sides of the whole window. Then, define a matrix of the form

```
sides <- matrix(data = c(left1, right1, bottom1, top1,

                         left2, right2, bottom2, top2,

                         ⋮

                         leftn, rightn, bottomn, topn),

          nrow = n, ncol = 4, byrow = TRUE)
```

The numbers assigned to `leftk`, `rightk`, `bottomk`, `topk` are as follows. For the `left` and `right` entries, numbers range from 0 (left-hand side of the active graphics window) to 1 (right-hand side of the graphics window).

Similarly, for `top` and `bottom`, numbers range from 0 (bottom of the graphics window) to 1 (top of the graphics window). Once the sides are defined, run

```
split.screen(figs = sides, erase = TRUE)
```

and, as before, to activate the kth screen for purposes of plotting, run

```
screen(n = k, new = TRUE)
```

All previous instructions remain the same.

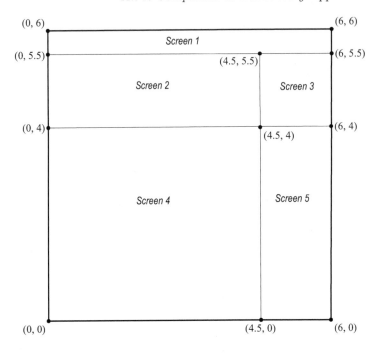

FIGURE 10.3
Partitioning of the 6 × 6 inch window used for Figure 10.1. In using the `split.screen` function, scale the coordinates by dividing the horizontal positions by the width of the full window, and the vertical positions by the height of the full window.

An Example

Consider using this approach to duplicate Figure 10.1. After opening a 6 × 6 inch graphics window using any one of the graphics device functions described in Section 7.6.1, set the partitioning side positions by running

```
> sides <- matrix(data = c(0.00, 1.00, 0.92, 1.00, #scrn1
+                          0.00, 0.75, 0.67, 0.92, #scrn2
+                          0.75, 1.00, 0.67, 0.92, #scrn3
+                          0.00, 0.75, 0.00, 0.67, #scrn4
+                          0.75, 1.00, 0.00, 0.67),#scrn5
+        nrow = 5, ncol = 4, byrow = TRUE)
```

This sets up the partitioning (almost) exactly as for Figure 10.1 in the manner described below and illustrated in Figure 10.3.

As shown in Figure 10.3, the top 1/2 inch of the 6×6 inch graphics window (screen 1) is reserved for the overall figure title. Since this occupies the full width of the window, it stretches from 0 to 1 horizontally. Since this screen occupies the top 1/2 inch of the full window, the vertical proportion it occupies is $0.5/6 \approx 0.08$, so this screen starts at 0.92 and ends at 1 vertically. Moving down to the next screen, which contains the horizontal boxplot, the horizontal starting point is 0 and its width is 4.5 inches. So, the horizontal proportion of the window it occupies is the first $4.5/6 = 0.75$ of the width. Vertically, this screen ranges from $4/6 \approx 0.67$ to $5.5/6 \approx 0.92$. Similarly, side positions as proportions of 1 are established for the remaining three screens.

To complete the Figure 10.1 duplication process, each plotting screen is then filled by first activating the desired screen using the `screen` function and then running the relevant code; see the script file for the complete code.

A Heads-up: It is important to not use any of these functions in combination with applications of `par(mfrow =c(x, y))`, `par(mfcol = (x, y))` or the `layout` function. Also, only occasionally, after using these functions a bit things may just stop working nicely; error messages start appearing and the `split.screen` function may not agree with anything given to it. This happens if something goes wrong with the way or the order in which the functions `screen` and `close.screen` are used. A first suggestion is to be very careful when preparing code for the use of these functions. If a problem arises, the simplest fix is to restart R; be sure to save any changes made to your script file before doing so.

10.2   Customizing Plotted Text and Symbols

Any number of setting changes can be made in a single `par` function call. Tables 10.1–10.4 list the arguments that are of interest here, and the script file contains exploratory code to illustrate how these arguments can be used for customizing what is plotted.

TABLE 10.1
Altering the point size and font family of plotted text. Note that pointsize, using `ps`, can be set only through calls of the `par` function.

`ps`	A positive integer, specifies the point size of plotted text only. The default is 12 point.
`family`	Examples include `"sans"` (Arial), `"serif"` (Times New Roman), and `"mono"` (Courier New). The default is `"sans"`.

TABLE 10.2

Scaling text and symbols. It is useful to remember that `cex` scales *all* plotted text and symbols relative to the default point-size.

`cex`	Global scaling of plotted text and symbols relative to the default. A 1 means no scaling, a positive number less than 1 results in a scale-down and a number greater than 1 results in a scale-up
`cex.axis`	Scaling tickmark labels relative to size specified by `cex`.
`cex.lab`	Scaling axes labels relative to the size specified by `cex`.
`cex.main`	Scaling the main title relative to the size specified by `cex`.
`cex.sub`	Scaling the subtitle relative to the size specified by `cex`.

TABLE 10.3

Altering the type-face of plotted text. These settings do not affect plotted symbols in any way, and also do not apply to all Hershey fonts.

`font`	Global type face for text. Includes 1 (plain text – default), 2 (bold face), 3 (italic), and 4 (bold face italic).
`font.axis`	Type face for axes tickmark label text.
`font.lab`	Type face for axes labels text.
`font.main`	Type face for main title text.
`font.sub`	Type face for sub title text.

TABLE 10.4

Specifying plotting colors to use; many color assignments can be made through character strings such as `"blue"`, `"green"`, `"brown"`, `"red"`, and so on.

`col`	Global plotting color within the current window, default is black.
`col.axis`	Color for axes tickmark labels.
`col.lab`	Color for axes labels.
`col.main`	Color for main title.
`col.sub`	Color for sub title.

With respect to Table 10.1, the default point size for text can be changed using `ps`, the default for plotting points remains at 12 points. In order to resize these the `cex` argument (see Table 10.2) should be used within the relevant plotting function. If this is done it *may* be necessary to balance the resulting change in text size through the `ps` argument.

It is helpful to remember that the new settings apply only to the current active plotting window. Once this window is closed, or if a new graphics window is opened, new plots will be constructed using the default settings. Also, if either of the `mfrow` or `mfcol` arguments is used to partition a plotting window, be sure to do all customizations *after* the partitioning is performed.

10.3   Inserting Mathematical Annotation in Plots

A small sampling of what can be done using fairly basic code appears in Figure 10.4. See the script file for the code, and see the documentation pages of `plotmath` for a detailed list of capabilities in this area. It will be noticed that the `expression` function serves the purpose of a "translator" for the `text` function when mathematical expressions are to be inserted in a figure.

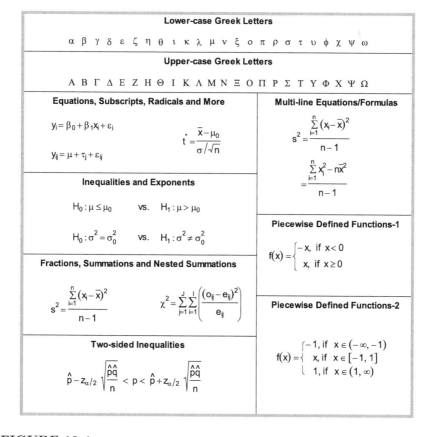

FIGURE 10.4
Plotting Greek letters and a variety of mathematical annotation.

Two other functions that may find use in this area are the `substitute` and `bquote` functions. For almost all cases of plotting mathematical annotation the `expression` function, enclosing the annotation desired, is assigned to any text plotting argument (such as `xlab`, `ylab`, `main`, or `sub`) in a plotting function. Annotation can also be included in legends by way of the `legend` function, or placed anywhere in a figure using the `text` function, or in margins of a figure using the `mtext` function.

Another useful function in this area is the `.()` function; here is a very simple example of its use with the `bquote` function. First generate a random sample,

```
> z <- sample(x = 1:100, size = 25, replace = TRUE)
```

then open a plotting window and include the axes for reference purposes. Next, plot various statistics centered at specified coordinates.

```
> text(x = 0, y = 0.5,
+    labels = bquote(bar(z) == .(mean(z))), adj = 0)
> text(x = 0.2, y = 0.6,
+    labels = bquote(s[z]^2 == .(var(z))), adj = 0)
> text(x = 0.4, y = 0.4,
+    labels = bquote(tilde(z) == .(median(z))), adj = 0)
> text(x = 0.6, y = 0.2,
+    labels = bquote(Range == .(
+       paste(range(z),collapse = " - "))), adj = 0)
```

The first application of the `text` function has the effect of writing in the value of $\bar{z}$ on the right-hand side of the equation `bar(z)==.(mean(z))`. As the next three applications of the `text` function suggest, the usefulness of the `.()` function lies in its capability to insert the value from a *single-valued* function evaluation directly in a figure.

10.4 More Low-Level Graphics Functions

There are a large variety of additional low-level graphics functions available in R for customizing plots. Some of these have already appeared in earlier illustrations of using other graphics features. Here a selection of such functions along with some examples of how they might be taken advantage of are provided. All of these functions need an active plotting window.

10.4.1 The points and symbols Functions

These functions operate in a manner fairly similar to previously encountered text function in that, given coordinates, the points and symbols functions can be used to plot objects to an existing plotting window.

The points Function

An earlier illustration of applying the points function appears in the code used to construct Figure 9.3, where the purpose was to add points to an existing plot. The usage definition of this function, with arguments of interest, is

```
points(x, y = NULL, type = "p", pch, cex, ...)
```

and the arguments serve the same purposes as for the plot function.

The symbols Function

The usage definition of this function, with arguments of interest, is

```
symbols(x, y = NULL, circles, squares, rectangles, stars,
    thermometers, boxplots, add = FALSE, fg = par("col"),
    bg = NA, inches = TRUE, ...)
```

One of six symbols (circles, squares, rectangles, stars, thermometers, or boxplots) can be plotted at coordinates assigned to x and y by including the desired symbol argument in the function call as described in the documentation pages for symbols. The color of the symbols and the fill-color can be assigned through the arguments fg and bg, respectively, as single values or vectors of values. The inches argument controls the size of the symbols plotted.

An illustration of using this function with the circles option to provide 3-dimensional information using a 2-dimensional plot is given in the script file.

10.4.2 The grid, segments, and arrows Functions

For all three of these functions the arguments col, lty, lwd and the notation "..." serve the same purposes as for previous line plotting graphics functions.

The grid Function

The usage definition of this function, with arguments of interest, is

```
grid(nx = NULL, ny = nx, col, lty, lwd)
```

Here is a description of the arguments: nx and ny represent the number of cells of the grid in the horizontal and vertical directions, respectively. Under the default setting nx = NULL results in the default assignment ny = nx, and the grid aligns with the tickmarks on the corresponding default axes with the tickmarks. If nx or ny is assigned the value NA, then no grid lines are drawn in the corresponding direction.

Code for three illustrations of plotting grids is given in the script for this section.

The segments Function

This function can be used to insert line segments from a collection of starting points to a corresponding collection of ending points. The function definition, along with arguments of interest, is

```
segments(x0, y0, x1, y1, col, lty, lwd, ...)
```

The arguments x0 and y0 are assigned coordinates for starting points, and the arguments x1 and y1 the coordinates of the corresponding end points for the line segments. To avoid technical difficulties, it is advisable to assign all of the arguments x0, y0, x1 and y1 vectors of the same length.

Figure 8.17 provides an illustration of how this function can be used to include a grid in a plot. As shown in the script for this section, one may create a more elaborate grid that identifies major and minor tickmarks, somewhat along the lines of the illustration provided in the sample code for the grid function.

The arrows Function

This function operates in exactly the same manner as the segments function, except in that it provides the additional feature of including an arrowhead at one or both ends of the plotted segment, referred to here as the shaft of the arrow.

The usage definition of this function, with arguments of interest, is

```
arrows(x0, y0, x1, y1, col, lty, lwd,
    length = 0.25, angle = 30, code = 2, ...)
```

The arguments x0, y0, x1 and y1 serve the same purpose as for the **segments** function. The argument length is used to fix the length of the arrowheads in inches and angle sets the angle between the shaft of the arrow and the edge of the arrow head. Both of these arguments are assigned a single value.

The integer assigned to code determines where the arrowhead is plotted along the shaft of the arrow. For example, with angle being assigned an acute angle: if code = 1 the arrowheads are drawn at, and pointing toward, the starting point coordinates; if code = 2 the arrowheads are drawn at, and pointing toward, the ending point coordinates; and if code = 3 the arrowheads are drawn at both the starting and ending point coordinates with the arrowheads pointing toward the starting and ending points, respectively. As with the arguments length and angle, code is assigned a single value.

10.4.3 Boxes, Rectangles, and Polygons

Here too the arguments col, lty, lwd and the notation "..." serve the same purposes as for previous line plotting graphics functions. The script for this section contains sample code for some examples of using these functions.

The box Function

The usage definition of this function, with arguments of interest, is

```
box(which = "plot", lty = "solid", lwd, col, ...)
```

The main argument here is which. One may choose to place a box around the plot (which = "plot") or around the whole figure (which = "figure"). Two other options available are outer and inner. These play a role in constructing boxes at inner and outer margin boundaries around figures containing multiple plots such as demonstrated in Figure 10.4. See the script file for code that provides an illustration of how these argument assignments can be taken advantage of.

The rect Function

The usage definition of this function, with arguments of interest, is

```
rect(xleft, ybottom, xright, ytop, col = NA,
    border = NULL, lty = par("lty"), lwd = par("lwd"), ...)
```

The horizontal positions of the sides of the rectangle(s) are provided through the arguments xleft and xright, and the vertical positions of the sides of the rectangle(s) are provided through ybottom and ytop. These arguments are assigned scalar (or vector) values. For this function col is used to set the color(s) with which the rectangles are to be filled, and may be assigned a single color or a vector of colors. Colors for rectangle border(s) are assigned through border, and border = NA is used to omit borders. As for the side position arguments, the arguments col, border, lty and lwd may also assigned single values or vectors of values (one for each rectangle plotted).

The polygon Function

The usage definition, with arguments of interest, is

```
polygon(x, y = NULL,
    border = NULL, col = NA, lty = par("lty"), ...)
```

The arguments x and y are assigned vectors containing the coordinates of the vertices of the polygon, and should look something like

```
x = c(x0, x1, ..., xn, x0), y = c(y0, y1, ..., yn, y0)
```

Then, line segments of type `lty` (default is a solid line), width `lwd` (default is 1), and `border` color (default is black, and `border = NA` omits a border) are drawn between adjacent vertices. The argument `col` is assigned the color to be used to fill the resulting polygon (the default is unfilled).

A simple example of using this function to provide a graphical representation for a cumulative probability distribution is given in the script for this section.

10.5 Error Bars

Informally, an error bar centered at a particular plotted point provides a feel for the spread of the observed data about some measure of central tendency. See, for example, [10], [32], [33], [85] and [88] for further discussions on this subject.

Here, the computational steps for obtaining bounds for error bars are summarized and then packaged in a function for use on specific types of data.

10.5.1 Computing Bounds for Error Bars

Error bars may be one of two broad classes. For descriptive error bars, the two common measures are the inter-quartile range and the sample standard deviation. Inferential error bars can be constructed using the standard error or a 95% confidence interval.

For $j = 1, 2, \ldots, n$ let $S_j = \{x_{1j}, x_{2j}, x_{3j}, \ldots, x_{n_j j}\}$ represent the jth observed sample. Then the bounds, $[L_j, U_j]$, for each sample's error bars can be obtained using any one of the following approaches.

Using a Range

One possibility is to use the interquartile range (from the first to the third quartile), in which case

$$L_j = Q_1(S_j) \quad \text{and} \quad U_j = Q_3(S_j).$$

These error bars include the middle 50% of the observations within each sample and are centered about the median of the sample.

Using the Sample Standard Deviation

First, for each $j = 1, 2, \ldots, n$, compute

$$\bar{x}_j = \frac{\sum_{i=1}^{n_j} x_{ij}}{n_j} \quad \text{and} \quad s_j^2 = \frac{\sum_{i=1}^{n_j} (x_{ij} - \bar{x}_j)^2}{n_j - 1}.$$

Then

$$L_j = \bar{x}_j - s_j \quad \text{and} \quad U_j = \bar{x}_j + s_j$$

identifies the spread within one standard deviation about the sample mean.

Using the Sample Standard Error

For each $j = 1, 2, \ldots, n$, compute

$$s_{\bar{x}_j} = \frac{s_j}{\sqrt{n_j}}.$$

Then

$$L_j = \bar{x}_j - s_{\bar{x}_j} \quad \text{and} \quad U_j = \bar{x}_j + s_{\bar{x}_j}$$

provides the spread within one standard error about the sample mean.

Using 95% Confidence Intervals

For each $j = 1, 2, \ldots, n$, first obtain t_j^* that satisfies

$$P(t \geq t_j^*) = 0.025$$

using the t-distribution with $\nu = n_j - 1$ degrees of freedom. Then

$$L_j = \bar{x}_j - t_j^* s_{\bar{x}_j} \quad \text{and} \quad U_j = \bar{x}_j + t_j^* s_{\bar{x}_j}.$$

provides the spread in the form of a 95% confidence interval centered at the sample mean.

10.5.2 The `errorBar.plot` Function

This function contains the computational steps just described, and is loaded in the workspace along with the other objects for this chapter. Code with detailed documentation can be viewed at the end of the script file for this chapter. The usage definition for this function is

```
errorBar.plot(y, g, type = "sd", sizes = TRUE,
    xlab = NULL, ylab = NULL, main = NULL, join = NULL,
    plot.stripchart = FALSE, month = FALSE,
    xlas = 1, ylas = 0)
```

The argument `y` is assigned a numeric vector whose entries are partitioned by the entries of the factor `g`, which may or may not be of an ordinal scale. The `type` of error bars may be chosen from one of four options: standard deviations, `"sd"` (the default); standard errors, `"se"`; 95% confidence intervals, `"ci"`; or interquartile ranges, `"iqr"`. If `sizes = TRUE` (the default), sample sizes for each level of the factor g are included in the plot. If `join = TRUE` is

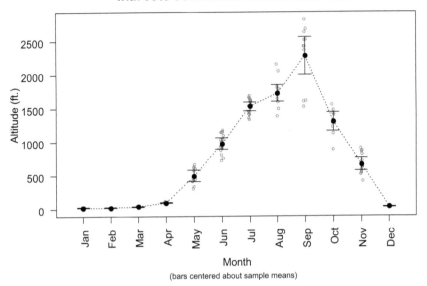

FIGURE 10.5

95% confidence interval error bars of altitudes for each month at which female deer were observed over the year 2009.

assigned, line segments are drawn between the centers of adjacent error bars, and the assignment `plot.stripchart = TRUE` results in the error bars being superimposed on the corresponding stripchart of the data. To have horizontal axis tickmarks be labeled by abbreviated month names set `month = TRUE`. Choosing 1 or 2 for `xlas` aligns the horizontal axis tickmark labels parallel or perpendicular to the horizontal axis, respectively, and choosing 0 or 1 for `ylas` aligns the vertical axis tickmark labels parallel or perpendicular to the vertical axis, respectively. The remaining arguments `xlab`, `ylab` and `main` are as before.

Figure 10.5 is the result of running the following code.

```
> with(data = subset(x = deerStudy,
+          subset = (gender == "Female") & (year == 2009)),
+     expr = errorBar.plot(y = altitude, g = month,
+          xlab = "Month", ylab = "Altitude (ft.)",
+          main = paste("2009 Female Deer Altitudes\n",
+               "with 95% Confidence Interval Error Bars"),
+          type = "ci", plot.stripchart = TRUE,
```

```
+          join = TRUE, sizes = FALSE, month = TRUE,
+          xlas = 2, ylas = 1))
```

Further sample applications of this function are contained in the script for this section.

10.5.3   Purpose and Interpretation of Error Bars

It is indicated in [10] that error bars are often used and interpreted incorrectly. For this reason, it is helpful to bring up some comments, suggestions and general rules of thumb in the uses and interpretations of error bars. The articles cited in the introductory paragraph of this section may be referred to for further details.

Purpose

The most obvious purpose of error bars is that they provide a pictorial display of the spread of the data in question in much the same way as stripcharts and boxplots do. The difference here is that error bars are focused about a specific point of interest. For example, the sample median or mean.

Some also use error bars for inferential purposes, specifically those constructed using standard errors or 95% confidence intervals. While this can be done, it is concerning such applications that much of the cautions in [10] address.

Interpretations and Some Suggestions

The simplest interpretations obtainable from error bars address the spread of the sample data, these are summarized in the following table.

Type	Basic Description
Inter-quartile Range	Spread of middle 50% of sample's values.
Standard Deviation	Spread of values within 1 standard deviation of the sample mean.
Standard Error	Spread of interval up to 1 standard error about a given sample's mean.
95% Confidence Interval	Spread of a 95% confidence interval.

To aid in the interpretation of error bars it is suggested in [32] that error bar figures should include a description of the type of error bar used, along with sample sizes and statistics used. It is also suggested that error bars should not be used for small sample sizes (for example, around 3). In such cases plots of the individual points should be preferred.

Cautions

Cautions outlined in [10] and [33] focus primarily on the use of error bars for inferential purposes and a useful discussion on some basic facts about interval estimates appears in [31]. Extreme caution is suggested when error bars are employed for purposes of pairwise comparisons, typically among population means. Firstly, only standard error and confidence interval error bars should be used; secondly, error bars should never be employed to perform pairwise comparisons using repeated measures on the same group.

<hr>

10.6 More R Graphics Resources

For those interested in expanding their graphics capabilities with R, there are many resources available on the internet and in print, including graphics-specific works such as, for example, [96] and [134]. All built-in graphics functions used in this book are from packages contained in the base subdirectories (part of the the basic/default download of R). Lewin-Koh provides a listing, along with details, of the many graphics packages available for R on the website

https://cran.r-project.org/web/views/Graphics.html

Of the many listed, the two packages that are gaining favor among users of R are `lattice` [117] and `ggplot2` [140]. User manuals for both of these packages can be found on the R Project website and websites devoted specifically to each of these packages, and the works [96] and [134] both provide introductions and examples.

11

Tests for One and Two Proportions

This chapter begins with a brief overview of using relevant built-in functions for computations involving probability distributions that will be needed in this chapter. The discussion then moves to constructing interval estimates (where appropriate) and testing hypotheses involving one and two population proportions based on observed frequencies associated with categorical variables. Code for all illustrations and figures, plus a bit more, are contained in the file Chapter11Script.R.

11.1 Relevant Probability Distributions

Four types of probability distributions find use in this chapter: binomial, hypergeometric, normal and chi-square distributions. This section provides an overview of the definitions and some relevant properties of each of these distributions along with their corresponding R functions.

11.1.1 Binomial Distributions

Let X be a *discrete random variable* that can acquire one of two values, 0 or 1, where 0 represents a failure (or false) and 1 represents a success (or true). Consider a *probability experiment* involving a single trial (for example, tossing a coin or rolling a die once), and denote the true probabilities of a success and a failure by p and $1 - p$, respectively. That is,

$$P(X = 1) = p \quad \text{and} \quad P(X = 0) = 1 - p.$$

The experiment described here is referred to as a *Bernoulli trial* and X is called a *Bernoulli random variable* with parameter p.

Now consider a fixed and finite number, n, of successive Bernoulli trials (for example, tossing a coin 10 times or rolling a die 23 times), the outcomes of each trial being independent of the outcomes of preceding trials. In this case let the random variable X represent the total number of successes that occur in the n trials. This probability experiment is called a *binomial experiment*, and in this case X is called a *binomial random variable*, having a *binomial distribution with parameters n and p*. Notation used to indicate this will be $X \sim BIN(n, p)$. The mean of X is $\mu = np$ and the variance is $\sigma^2 = np(1-p)$.

The *sample space* for a binomial experiment involving n trials is $\{0, 1, 2, \ldots, n\}$, and the probability of obtaining exactly x successes is given by the *probability density function*[1]

$$P(X = x) = \binom{n}{x} p^x (1-p)^{n-x}.$$

The R function corresponding to this density function, with arguments of interest, is

```
dbinom(x, size, prob)
```

where `size` is assigned the number of trials, n, and `prob` the probability of a success, p. The argument `x` can be assigned a single value or a vector of values from the sample space.

Cumulative probabilities associated with the binomial distribution are obtained using the *cumulative distribution function*

$$P(X \le x) = \sum_{k=0}^{x} \binom{n}{k} p^k (1-p)^{n-k}.$$

Computing cumulative probabilities with R involves using the function

```
pbinom(q, size, prob, lower.tail = TRUE)
```

where the arguments `size` and `prob` are the same as described previously. While `q` may be assigned a vector of integers from the sample space, $\{0, 1, 2, \ldots, n\}$, most applications in this book will involve computing a single probability of the form $P(X \le x)$. In such cases the *quantile* argument `q` is assigned whatever value x is given. If `lower.tail = TRUE` (the default setting) the sum of probabilities from $X = 0$ up through $X = x$ is computed. To compute $P(X > x)$, set `lower.tail = FALSE`.

A third function available for computations involving a binomial distribution is

```
qbinom(p, size, prob, lower.tail = TRUE)
```

---

[1]The binomial coefficient $\binom{n}{x}$, alternatively denoted by nC_x, is defined by

$$\binom{n}{x} = \frac{n!}{(n-x)!\,x!},$$

where $n! = n(n-1)\cdots 3 \cdot 2 \cdot 1$ (read as n factorial) and $0! = 1$. The function `factorial(x)` computes $x!$ and the function `choose(n,k)` computes $\binom{n}{k}$.

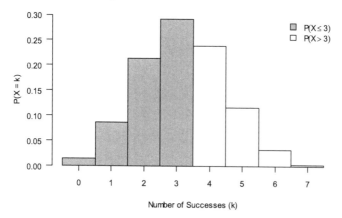

FIGURE 11.1
Density plot of $X \sim BIN(7, 0.45)$. Since X is a discrete random variable and the width of the bars equal 1, the area of each rectangle equals its height and, consequently, heights of the bars represent probabilities. Cumulative probabilities are obtained by summing the areas of the bars in question.

The argument p represents a probability or a vector of probabilities for which corresponding quantiles are desired. This function can be used in two ways. Let P represent a given cumulative probability. The first application, using `lower.tail = TRUE`, involves finding the smallest nonnegative integer x for which $P(X \leq x) \geq P$. The second application, using `lower.tail = FALSE`, involves finding the smallest nonnegative integer x for which $P(X > x) < P$.

Examples

Let $X \sim BIN(7, 0.45)$, and refer to Figure 11.1 for the following. This plot was constructed with the help of the functions `dbinom` and `barplot`. Running

```
> dbinom(x = 5, size = 7, prob = 0.45)
```

gives $P(X = 5) = 0.1172215$, and to calculate probabilities $P(X = x)$ for each $x = 0, 1, \ldots, 7$ run

```
> dbinom(x = 0:7, size = 7, prob = 0.45)
```

Next, $P(X \leq 3) = 0.6082878$ is found by running

```
> pbinom(q = 3, size = 7, prob = 0.45, lower.tail = TRUE)
```

and to calculate $P(X > 3)$ run

```
> pbinom(q = 3, size = 7, prob = 0.45, lower.tail = FALSE)
```

If a right-tailed probability of the form $P(X \geq 3)$ is needed, run

```
> pbinom(q = 2, size = 7, prob = 0.45, lower.tail = FALSE)
```

Dropping down to 2 ensures that $P(X = 3)$ is captured in the computation.

On occasion it may be that a probability of the form $P(3 \leq X \leq 5)$ may be desired. To get this run

```
> pbinom(q = 5, size = 7, prob = 0.45) -
+      pbinom(q = 2, size = 7, prob = 0.45)
```

Note that the argument `lower.tail` is left in its default setting. Alternative and shorter code involving the `sum` function and the function `dbinom`, or the `diff` function and the function `pbinom` can be written to calculate this quantity; see the script file.

Now consider finding the smallest nonnegative integer x for which the probability $P(X \leq x) \geq 0.65$ holds true. The answer can be found by running either of the lines

```
> pbinom(q = 0:7, size = 7, prob = 0.45, lower.tail = TRUE)
> qbinom(p = 0.65, size = 7, prob = 0.45,
+      lower.tail = TRUE)
```

It is found that the smallest nonnegative integer x for which $P(X \leq x) \geq 0.65$ is $x = 4$. Similarly, the smallest nonnegative integer x for which $P(X > x) < 0.15$ can be found using either of the lines

```
> pbinom(q = 0:7, size = 7, prob = 0.45,
+      lower.tail = FALSE)
> qbinom(p = 0.15, size = 7, prob = 0.45,
+      lower.tail = FALSE)
```

In this case $x = 5$. Unlike for continuous random variables, binomial probabilities for individual values of X in the sample space are nonzero. For this reason it is important to pay attention to whether or not a particular value of X should be included in computations involving a cumulative probability.

11.1.2   Hypergeometric Distributions

Let a population contain a total of N objects or subjects, m of one type and $N - m$ of another type. Suppose k of these N objects or subjects are randomly selected without replacement resulting in $X = x$ of those selected being of the first type. Then the discrete random variable X is said to have a *hypergeometric distribution with parameters k, m and N*. Notation used to indicate this will be $X \sim HYP(k, m, N)$, and the mean and variance of such a distribution are

$$\mu = km/N \quad \text{and} \quad \sigma^2 = km \left(1 - \frac{m}{N}\right) \left(\frac{N-k}{N-1}\right) \bigg/ N,$$

respectively. The probability density function for this discrete distribution is

$$P(X = x) = \binom{m}{x} \binom{N-m}{k-x} \bigg/ \binom{N}{k},$$

where, for non-zero probabilities, the sample space contains the integers $\max(0, k - N + m) \leq x \leq \min(k, m)$ and with possible values of k being $k = 1, 2, \ldots, N$ and of m being $m = 0, 1, \ldots, N$.

The R function for this density function with arguments of interest is

```
dhyper(x, m, n, k)
```

Here m represents the number of Type 1 objects in the population/collection, n the number of Type 2 objects (that is, n $=N-m$ in the above notation), and x is the number of Type 1 objects in k randomly selected objects. Cumulative probabilities are given by

$$P(X \leq x) = \sum_{j=a}^{x} \binom{m}{j} \binom{N-m}{k-j} \bigg/ \binom{N}{k},$$

with $a = \max(0, k - N + m)$, and computed in R using

```
phyper(q, m, n, k, lower.tail = TRUE)
```

where q is the quantile up to which the probability is to be computed if lower.tail = TRUE and the remaining arguments are as previously defined. Given probabilities, p, the function

```
qhyper(p, m, n, k, lower.tail = TRUE)
```

is used to compute hypergeometric quantiles in much the same manner as with the function qbinom.

Examples

Let $X \sim HYP(17, 7, 23)$, and refer to Figure 11.2 for the following. For convenience, note that $k = 17$, $m = 7$ and $n = N - m = 16$, so the allowable values for x are $1 \leq x \leq 7$ (all other values result in a zero probability). Running

```
> dhyper(x = 1, m = 7, n = 16, k = 17)
```

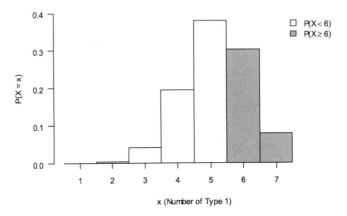

FIGURE 11.2
Density plot of the random variable $X \sim HYP(17, 7, 23)$. The horizontal axis
here represents the number of observations of Type 1 objects selected. Desired
cumulative probabilities are obtained as for the binomial distribution.

yields $P(X = 1) = 6.934332e-05$. This is a very small number, which explains
an apparent absence of a bar at $X = 1$ in Figure 11.2. Run

```
> phyper(q = 5, m = 7, n = 16, k = 17, lower.tail = TRUE)
```

to compute $P(X < 6) = P(X \leq 5)$, and to compute $P(X \geq 6)$ run

```
> phyper(q = 5, m = 7, n = 16, k = 17, lower.tail = FALSE)
```

The suggestions provided for computing binomial cumulative probabilities are
also applicable to this discrete probability distribution.
    To find the smallest nonnegative integer x for which $P(X \leq x) \geq 0.05$,
use

```
> qhyper(p = 0.05, m = 7, n = 16, k = 17,
+        lower.tail = TRUE)
```

to get $x = 4$. Similarly, to find the smallest nonnegative integer x for which
$P(X > x) < 0.05$, the code

```
> qhyper(p = 0.05, m = 7, n = 16, k = 17,
+        lower.tail = FALSE)
```

yields $x = 7$.

11.1.3   Normal Distributions

Let X be a *continuous random variable* for which the sample space is the set of all real numbers, $(-\infty, \infty)$. Denote the mean of X by μ and the variance by σ^2, and define the probability density function of X by

$$f(X) = \frac{1}{\sqrt{2\pi}\,\sigma} e^{-\frac{(X-\mu)^2}{2\sigma^2}}.$$

Then X is said to be *normally distributed with parameters μ and σ*. Notation used to indicate this will be $X \sim N(\mu, \sigma)$.

The R function corresponding to this probability density is

```
dnorm(x, mean = 0, sd = 1)
```

The default parameter settings for this function is for the *standard normal* (or *Gaussian*) *distribution* ($\mu = 0$ and $\sigma = 1$). As for previous R density functions, the argument x can be assigned individual values or a vector of values.

Since X is a continuous random variable, the cumulative distribution function is defined by an integral,[2]

$$P(X \le x) = \int_{-\infty}^{x} f(t)\,dt.$$

Also, for individual points x in the sample space $P(X = x) = 0$, so unlike in computations for discrete probability distributions,

$$P(X \le x) = P(X < x) \quad \text{and} \quad P(X \ge x) = P(X > x).$$

The R function for computing this integral for a given $X = x$ is

```
pnorm(q, mean = 0, sd = 1, lower.tail = TRUE)
```

The arguments q, mean and sd are as before, and lower.tail = TRUE instructs R to compute a left-tailed probability. To compute $P(X > x)$, set lower.tail = FALSE.

As for previous quantile functions, the function,

```
qnorm(p, mean = 0, sd = 1, lower.tail = TRUE)
```

returns the approximate quantile corresponding to a given left- or right-tailed probability, p.

Examples

The graph of the density function for $X \sim N(3.1, 1.2)$ in Figure 11.3 was constructed using the dnorm function within the curve function. For continuous random variables this is the only way probability density ("d") functions such as dnorm will be used in this and in later chapters. To calculate the left-tailed probability $P(X \le 4.5)$, run

---

[2]R performs all necessary computations so, while helpful, a knowledge of calculus is not needed for this book.

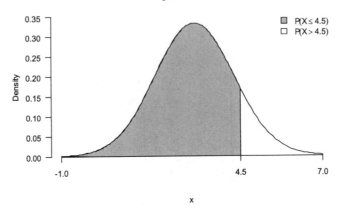

FIGURE 11.3
Density curve of $X \sim N(3.1, 1.2)$. The area of the shaded region under the curve represents a left-tailed probability and the area of the unshaded region a right-tailed probability.

```
> pnorm(q = 4.5, mean = 3.1, sd = 1.2, lower.tail = TRUE)
```

Alternatively, to calculate the right-tailed probability $P(X > 4.5)$, run

```
> pnorm(q = 4.5, mean = 3.1, sd = 1.2, lower.tail = FALSE)
```

To calculate the probability $P(2 \leq X \leq 5)$, run

```
> pnorm(q = 5, mean = 3.1, sd = 1.2) -
+     pnorm(q = 2, mean = 3.1, sd = 1.2)
```

Note that the default setting `lower.tail = TRUE` is used, so this need not be specified in the two `pnorm` function calls. Again, the `diff` function can be used here to shorten the code for this calculation; see the script file.

Now consider approximating the quantile x for which $P(X \leq x) = 0.05$. Since $P(X \leq x)$ represents a left-tailed probability, run

```
> qnorm(p = 0.05, mean = 3.1, sd = 1.2, lower.tail = TRUE)
```

Alternatively, run

```
> qnorm(p = 0.05, mean = 3.1, sd = 1.2, lower.tail = FALSE)
```

to find x for which $P(X > x) = 0.05$ since this computation involves a right-tailed probability.

11.1.4 Chi-Square Distributions

Let X be a continuous random variable for which the sample space is the set of all positive real numbers, $(0, \infty)$. Let ν represent any positive integer, and define the probability density function of X by[3]

$$f(X) = \frac{1}{2^{\nu/2}\, \Gamma\,[\nu/2]} x^{\nu/2-1}\, e^{-X/2}.$$

Then X is said to have a *chi-square distribution with parameter* ν.[4] Notation used to indicate this will be $X \sim \chi^2(\nu)$. The mean for this distribution is $\mu = \nu$ and the variance is $\sigma^2 = 2\nu$.

The R function corresponding to this probability density function, with arguments of interest, is

```
dchisq(x, df)
```

Again, the argument x can be assigned individual values or a vector of values and df is assigned the degrees of freedom, ν, of the distribution in question.

Since X is a continuous random variable, the cumulative distribution function is defined by an integral,

$$P(X \le x) = \int_0^x f(t)dt,$$

and the R function for computing this integral for a given $X = x$ is

```
pchisq(q, df, lower.tail = TRUE)
```

The arguments q, df and lower.tail are used as described previously. As for normal probability distributions, there may be a need to find solutions to equations of the form $P(X \le x) = P$ or $P(X > x) = P$. As might be expected, the function

---

[3] The function $\Gamma\,[\kappa]$ that appears in this probability density function is called the *gamma function*, and is defined by

$$\Gamma\,[\kappa] = \int_0^\infty t^{\kappa-1} e^{-t} dt,$$

where $\kappa > 0$. The R function to compute this is

```
gamma(x)
```

where x is assigned the value κ.

An understanding of the properties of this function is not necessary here. The interested reader is referred to works such as [7, p. 111] for further details on this function and the associated *gamma distribution*; see also ?GammaDist.

[4] The parameter ν is also referred to as the *degrees of freedom* and is defined in most texts to be the number of observations in a sample whose values can vary, yet still yield the same computed statistic. The concept of degrees of freedom is specific to sampling distributions, which find use in hypothesis testing. See, for example, [62] or [102] for further discussions on this term.

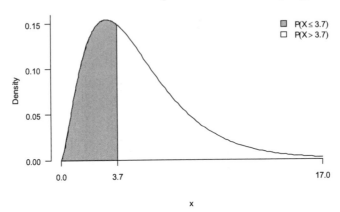

FIGURE 11.4
Density plot of $X \sim \chi^2(5)$. As before, the shaded area under the curve represents a left-tailed probability and the unshaded area a right-tailed probability.

```
qchisq(p, df, lower.tail = TRUE)
```

is used. The arguments for this function are as defined previously.

Examples

Let $X \sim \chi^2(5)$, the density curve of which is shown in Figure 11.4. To calculate the left-tailed probability $P(X \le 3.7)$ and the right-tailed probability $P(X > 3.7)$, run the lines

```
> pchisq(q = 3.7, df = 5, lower.tail = TRUE)
> pchisq(q = 3.7, df = 5, lower.tail = FALSE)
```

and to calculate $P(3 \le X \le 10)$, run

```
> pchisq(q = 10, df = 5) - pchisq(q = 3, df = 5)
```

Once again, the `diff` function can be used to shorten this code. Finally, to find x such that $P(X \le x) = 0.05$, run

```
> qchisq(p = 0.05, df = 5, lower.tail = TRUE)
```

and run

```
> qchisq(p = 0.05, df = 5, lower.tail = FALSE)
```

to find x such that $P(X > x) = 0.05$.

11.2 Single Population Proportions

Let $X \sim BIN(n, p)$, with n being the number of trials and p the probability of a success (or population proportion) in any trial and, for purposes of this section, denote the probability of a failure by $q = 1 - p$. The tasks for this section include methods used to estimate the population proportion p, and tests involving a hypothesized population proportion, p_0.

11.2.1 Estimating a Population Proportion

If n is large enough to ensure that both $np \geq 5$ and $nq \geq 5$ hold true (referred to as the *nonskewness criterion*), then X can be approximated by the normal random variable with mean $\mu = np$ and standard deviation $\sigma = \sqrt{npq}$. This is a consequence of the Central Limit Theorem, see Section 11.4.1 for a simulation that illustrates this result graphically.

Now suppose n trials result in x successes, then the sample proportion of successes is $\hat{p} = x/n$. Denote the proportion of failures by $\hat{q} = 1 - \hat{p}$, then if n is sufficiently large so that both $np \geq 5$ and $nq \geq 5$ hold true[5] the previously mentioned result implies that the distribution of $\hat{p}$ is approximately normally distributed with mean $\mu_{\hat{p}} = p$ and standard deviation $\sigma_{\hat{p}} = \sqrt{pq/n}$. Using this approximation, the $100(1 - \alpha)\%$ confidence interval for p is given by[6]

$$\max\left\{0, \hat{p} - z_{\alpha/2}\sqrt{\frac{\hat{p}\hat{q}}{n}}\right\} < p < \min\left\{\hat{p} + z_{\alpha/2}\sqrt{\frac{\hat{p}\hat{q}}{n}}, 1\right\}$$

with $z_{\alpha/2}$ satisfying the right-tailed probability $P(Z \geq z_{\alpha/2}) = \alpha/2$ where $Z \sim N(0, 1)$.

Solving the expression for the *maximum error of estimate*,

$$E = z_{\alpha/2}\sqrt{\frac{\hat{p}\hat{q}}{n}},$$

for n and rounding up to the nearest integer yields

$$n \approx \left\lceil \bar{p}\bar{q}\left(\frac{z_{\alpha/2}}{E}\right)^2 \right\rceil.$$

This makes it possible to determine what sample size would ensure a desired

---

[5]The sample proportion of successes and failures can be used as approximate check of these two conditions; that is, to determine if $n\hat{p} \geq 5$ and $n\hat{q} \geq 5$ are both true.

[6]Note that if 0 is chosen for the lower bound, or 1 for the upper bound the inequality $<$ is replaced by $\leq$.

Aside from specific scenarios such as these, all confidence intervals in this book will be expressed as open intervals; in the literature, the preference of which convention to use, open versus closed intervals, may vary from author to author.

error for a given confidence level before data collection is begun. The value used for $\bar{p}$ may be either a prior sample estimate or, if a prior estimate is unavailable, $\bar{p} = 0.5$ is used.

See [97] for examples of alternative methods that may be used to obtain two-sided confidence intervals for a single population proportion.

Example

An Elementary Statistics student wished to estimate the proportion of Juneau residents whose standard footwear is the ubiquitous Extra Tuff boot. She decided on a 90% confidence level, and wanted a maximum error of estimate of at most $E = 0.05$. She surveyed 275 randomly selected Juneau residents and found that 173 of them wore Extra Tuffs for their everyday activities.

It is a simple matter to verify that her sample is large enough, and code can be prepared from scratch to compute the interval as described above (see script file). The result shows that a 95% confidence interval for the true proportion of Juneau residents who wear Extra Tuffs for their everyday activities is approximately $(0.581, 0.677)$.

Alternatively, the `prop.test` function can be used to obtain such an interval. However, the manner in which the interval bounds are computed differs slightly from above.[7] Because of this the bounds produced by `prop.test` differ from bounds computed using the formula presented in the introductory comments, particularly for cases where $\hat{p}$ is closer to 0 or 1 for samples of sizes less than 1000; see the script file for some exploratory code.

11.2.2 Hypotheses for Single Proportion Tests

If p_0 is the hypothesized population proportion, where typically $0 < p_0 < 1$, possible alternative hypotheses associated with such tests are

$$H_1 : p < p_0, \qquad H_1 : p \neq p_0, \qquad H_1 : p > p_0.$$

Example

On a recent visit to the Pack Creek Bear Viewing Area near Juneau, a tourist was informed by the accompanying (newly hired) intern that adult

---

[7]Without a correction for continuity, the interval bounds are computed using

$$\frac{\hat{p} + z_{\alpha/2}^2/(2n) \pm z_{\alpha/2} \sqrt{\hat{p}\hat{q}/n + z_{\alpha/2}^2/(4n^2)}}{1 + z_{\alpha/2}^2/n}.$$

In [97] this method is referred to as the *Wilson Score method*. Observe that as the sample size n increases this interval formula produces bounds that approach those just computed using what is referred to as the *Wald method* in [97]. See this article for a comparison of these two and other methods for single proportion confidence intervals.

male brown bears make up less than 30% of the bears that are seen there. Unknown to the intern, recent randomly gathered field data from the viewing area found that of the 35 distinct bears seen, there were 7 adult male brown bears.

Let p denote the true, but unknown, proportion of distinct adult brown bears that visit the Pack Creek Bear Viewing Area. Then, the hypotheses corresponding to the intern's claim are

$$H_0 : p \geq 0.30 \qquad \text{vs.} \qquad H_1 : p < 0.30 \quad \text{(Claim)}$$

Three approaches to testing these hypotheses are described here.

11.2.3 A Normal Approximation Test

This test is commonly referred to as the large-sample z-test for a single population proportion. Let x be the number of successes observed in n trials, $\hat{p}$ the resulting observed proportion of successes, and p_0 the hypothesized proportion of successes. If both $np_0 \geq 5$ and $nq_0 \geq 5$ hold true then, under the null hypothesis $(p = p_0)$, $\hat{p}$ may be assumed to be approximately normally distributed with mean $\mu_{\hat{p}} = p$ (the true proportion of successes) and standard deviation $\sigma_{\hat{p}} = \sqrt{pq/n}$. If n is sufficiently large the test statistic is computed using

$$z^* = \frac{\hat{p} - p_0}{\sqrt{p_0 q_0 / n}},$$

and, under the null hypothesis, $z^* \sim N(0,1)$.

Since a continuous distribution is being used to approximate a discrete distribution, there are occasions when a *correction for continuity* of $\varepsilon = 1/(2n)$ is recommended. It is worth noting, however, that including the correction for continuity results in a more conservative test (see, for example, [123, p. 255]) and that the size of this correction decreases as the sample size increases. Additionally, the correction for continuity is used only if $\varepsilon \leq |\hat{p} - p_0|$.

To employ the correction for continuity, using the notation

$$\delta = \begin{cases} 1 & \text{for left-tailed probabilities, and} \\ -1 & \text{for right-tailed probabilities,} \end{cases}$$

the test statistic is computed using

$$z^* = \frac{\hat{p} - p_0 - \delta \varepsilon}{\sqrt{p_0 q_0 / n}}.$$

In either case, whether or not the correction for continuity is employed, the p-value for this test is obtained using $Z \sim N(0,1)$ according to the conventions:

$$\text{p-value} = \begin{cases} P(Z \leq z^*) & \text{for a left-tailed test;} \\ P(Z \geq z^*) & \text{for a right-tailed test.} \end{cases}$$

and, for a two-tailed test,

$$\text{p-value} = \begin{cases} 2P(Z \le z^*) & \text{if } P(Z \le z^*) \le 0.5; \\ 2P(Z \ge z^*) & \text{if } P(Z \le z^*) > 0.5. \end{cases}$$

The null hypothesis is then rejected if p-value $\le \alpha$.[8]

Example

For the bear viewing example in question $p_0 = 0.30$, $x = 7$ and $n = 35$, and once again code to perform the test can be prepared from scratch. It is found that $z^* = -1.291$ and p-value $= 0.09835$. So, unless a significance level of $\alpha \ge 0.09835$ is chosen, there is insufficient evidence to support the intern's claim that adult male brown bears make up less than 30% of the bears that are seen at the Pack Creek Bear Viewing Area. The use of the `prop.test` function in performing an equivalent test is now demonstrated.

11.2.4 A Chi-Square Test

The strategy for this procedure, which is equivalent to the previous z-test and is due to Pearson, is to follow the approach described later in this chapter for performing tests associated with contingency tables. As before, let x represent the number of successes observed in n trials. Then, using the notation $o_1 = x$ and $o_2 = n - x$ for *observed frequencies*, and $e_1 = p_0\,n$ and $e_2 = q_0\,n$ for *expected frequencies* (where p_0 and q_0 are as defined previously), the test statistic is computed using

$$\chi^{2*} = \sum_{i=1}^{2} \frac{(o_i - e_i)^2}{e_i}.$$

If $e_1 \ge 5$ and $e_2 \ge 5$ both hold true then under the null hypothesis $\chi^{2*} \sim \chi^2(1)$ and the p-value is computed as follows.[9]

For a two-tailed test the p-value is computed using the chi-square distribution with $\nu = 1$ degree of freedom,

$$\text{p-value} = P(\chi^2 \ge \chi^{2*}).$$

For a one-tailed test, first compute

$$z^* = \sqrt{\chi^{2*}},$$

---

[8]The rejection criterion p-value $\le \alpha$ is used because confidence intervals are expressed as open intervals. Note that it is extremely unlikely that p-value $= \alpha$ will occur.

[9]There are two observed frequencies for such cases. For a computed statistic to remain constant, one observed frequency is fixed and the remaining observation is allowed to vary. So, the Chi-square distribution associated with the statistic χ^{2*} has $\nu = 1$ degree of freedom.

then use $Z \sim N(0,1)$ to compute the right-tailed probability[10]

$$\text{p-value} = P(Z \geq z^*).$$

The function for this approach is the earlier mentioned `prop.test`, and the general usage definition is

```
prop.test(x, n, p,
          alternative, conf.level, correct = TRUE)
```

This is a versatile function which can be used to test a variety of hypotheses associated with proportions. However, for the present purpose `x` is assigned the number of observed successes, x, `n` is assigned the number of trials, n, and `p` is assigned the hypothesized proportion p_0. The argument `alternative` is used to identify the test as left-tailed (`alternative = "less"`), two-tailed (`alternative = "two.sided"`), or right-tailed (`alternative = "greater"`). The default setting for the confidence level is 95% (`conf.level = 0.95`). The value assigned to this argument is used in computing the confidence interval that corresponds to the test performed. The default setting for a correction for continuity is `correct = TRUE`.

Example

A p-value identical to that of the previous method comes from running

```
> prop.test(x = 7, n = 35, p = 0.30, alternative = "less",
+      conf.level = .95, correct = FALSE)
```

The *correction for continuity* is not used here,[11] hence `correct = FALSE`. In this case the output gives $\chi^{2*} = 1.6667$, $df = 1$ and p-value $= 0.09835$ and the inference is as for the previous example.

A Comment about the Confidence Interval Produced: The 95% confidence interval produced by `prop.test` for this left-tailed test, to three decimal places, is

$$0 \leq p < 0.331.$$

---

[10]This is how the `prop.test` function goes about one-tailed tests for this approach. Indeed, some algebra shows that $(z^*)^2 = \chi^{2*}$; moreover, under the null hypothesis $(z^*)^2$ has a χ^2-distribution with $\nu = 1$ degree of freedom – see, for example, [7, p. 271]. In fact, if a correction for continuity is not used, the normal approximation (z-test) and this approach are equivalent.

[11]Some suggest a correction for continuity is not necessary or should not be used, others suggest its use under certain conditions. Regardless, it is generally agreed that if the correction for continuity is employed the resulting test is conservative (less likely to result in a rejection of a false null hypothesis).

According to the details provided for the function `prop.test`, the default setting is `correct = TRUE` and under this setting (as applicable to a single proportion test) the correction $\varepsilon = 1/(2n)$ is applied only if $\varepsilon \leq |\hat{p} - p_0|$.

The left-bound is set at 0 since a proportion cannot be less than 0. The hypotheses stated can be equivalently tested as follows. Since the hypothesized value, $p_0 = 0.3$, lies inside the confidence interval, at the $\alpha = 0.05$ level of significance the null hypothesis is not rejected.

11.2.5 An Exact Test

The p-value can be calculated directly from the binomial distribution with parameters n and p_0 as follows: For one-tailed tests, use

$$\text{p-value} = \begin{cases} P(X \leq x) & \text{for left-tailed tests; and} \\ P(X \geq x) & \text{for right-tailed tests.} \end{cases}$$

For two-tailed tests, if x is very *close* to np_0 in value (up to some desired tolerance) set p-value $= 1$. Otherwise, first compute $m = \lfloor np_0 \rfloor$ (that is, round np_0 *down* to the nearest non-negative integer), then assuming $x \neq m$, use[12]

$$\text{p-value} = \begin{cases} P(X \leq 2m - x) + P(X > x) & \text{if } x > np_0; \text{ and} \\ P(X \leq x) + P(X > 2m - x) & \text{if } x < np_0. \end{cases}$$

The function for this approach, along with arguments of interest, is

```
binom.test(x, n, p, alternative, conf.level)
```

where the arguments included serve the same purposes as for the `prop.test` function.

Example

This test is conservative, resulting in a larger p-value than the previous methods. Running

```
> binom.test(x = 7, n = 35, p = 0.30, alternative = "less",
+      conf.level = .95)
```

produces $\hat{p} = 0.2$, p-value $= 0.1326$. Yet again, there is insufficient evidence to support the intern's claim that adult male brown bears make up less than 30% of the bears that are seen at the Pack Creek Bear Viewing Area. It will be noticed that the confidence interval corresponding to this method is wider than that produced by the previous method.

---

[12]The applicable R function, `binom.test`, employs an algorithm that amounts to using a tolerance that varies from *around* 0.1 to 0.5, depending on the value of $P(X = x)$. The strategy described here produces identical results, except when x is very close to (but not equal to) np_0, see [70, pp. 20 and 24].

11.2.6 Which Approach Should Be Used?

If the sample size, n, is sufficiently large to ensure that the conditions for the normal approximation of the binomial distribution with parameters n and p_0 are satisfied, any one of the above procedures will suffice. However, as mentioned previously, the exact approach is more conservative. Otherwise, the exact approach should be used with the understanding that this is a conservative test.

11.3 Two Population Proportions

Let $X_1 \sim BIN(n_1, p_1)$ and $X_2 \sim BIN(n_2, p_2)$ represent two independent binomially distributed random variables The tasks for this section include methods used to estimate the difference between the two population proportions, and to test a hypothesized difference of the form $p_1 - p_2 = p_0$, the hypothesized difference p_0 typically being zero.

11.3.1 Estimating Differences Between Proportions

Let x_1 be the number of observed successes from n_1 trials on one population and x_2 the number of observed successes from n_2 trials on a second population. Denote the sample proportions of successes by $\hat{p}_1 = x_1/n_1$ and $\hat{p}_2 = x_2/n_2$, and of failures by $\hat{q}_1 = 1 - \hat{p}_1$ and $\hat{q}_2 = 1 - \hat{p}_2$.

Now, if the two samples are independent of each other, and n_1 and n_2 are sufficiently large to ensure (as a general rule of thumb) that: Either it is known that $n_1 p_1 \geq 5$, $n_1 q_1 \geq 5$, $n_2 p_2 \geq 5$, and $n_2 q_2 \geq 5$ all hold true; or based on the observed values, $n_1 \hat{p}_1 \geq 5$, $n_1 \hat{q}_1 \geq 5$, $n_2 \hat{p}_2 \geq 5$, and $n_2 \hat{q}_2 \geq 5$ all hold true, then, as for the single population proportion case, the difference $p_1 - p_2$ can be estimated using a normal approximation. That is,

$$\max\left\{-1, (\hat{p}_1 - \hat{p}_2) - z_{\alpha/2}\sqrt{\frac{\hat{p}_1\hat{q}_1}{n_1} + \frac{\hat{p}_2\hat{q}_2}{n_2}}\right\}$$

$$< p_1 - p_2 <$$

$$\min\left\{1, (\hat{p}_1 - \hat{p}_2) + z_{\alpha/2}\sqrt{\frac{\hat{p}_1\hat{q}_1}{n_1} + \frac{\hat{p}_2\hat{q}_2}{n_2}}\right\}$$

with $z_{\alpha/2}$ satisfying $P(Z \geq z_{\alpha/2}) = \alpha/2$ with $Z \sim N(0, 1)$.

Example

Suppose field observations over a five-year period yielded 153 distinct bear sightings in the Sitka area, 93 of which were black bears. For the same period, suppose that 103 distinct bear sightings in the Juneau area included 73 black bears.

Consider obtaining a 90% confidence interval to estimate the difference between the true proportions of black bears in the Sitka and Juneau areas. Place the data in a matrix, `bears`, and then run, for example,

```
> round(prop.test(x = bears["Black",], n = bears["Total",],
+           alternative = "two.sided", correct = FALSE,
+           conf.level = 0.90)$conf.int[1:2], digits = 4)
```

to get $-0.1991 < p_1 - p_2 < -0.0027$. See the script file for some further exploratory code with the `prop.test` function using these data.

11.3.2 Hypotheses for Two Proportions Tests

If p_1 and p_2 represent the true proportions for the populations, possible alternative hypotheses associated with such tests include[13]

$$H_1 : p_1 < p_2, \qquad H_1 : p_1 \neq p_2, \qquad H_1 : p_1 > p_2.$$

Example

Continuing with the previous bears example, consider testing the claim that there is a higher proportion of brown bears in the Sitka area than in the Juneau area.

Let p_S represent the true proportion of brown bears in the Sitka area, and p_J the proportion in the Juneau area. Then the hypotheses to be tested are

$$H_0 : p_S \leq p_J \qquad \text{vs.} \qquad H_1 : p_S > p_J \quad \text{(Claim)}$$

Two equivalent approximate approaches, and an exact method for testing such hypotheses are presented here.

11.3.3 A Normal Approximation Test

This test is commonly referred to as the large-samples z-test for two population proportions. Assume the two samples are independent of each other and suppose that n_1 and n_2 are sufficiently large to ensure that $n_1 p_1 \geq 5$, $n_1 q_1 \geq 5$, $n_2 p_2 \geq 5$, and $n_2 q_2 \geq 5$ all hold true.[14] Then the test statistic is

---

[13] Another way to state these alternatives is

$$H_1 : p_1 - p_2 < p_0, \qquad H_1 : p_1 - p_2 \neq p_0, \qquad H_1 : p_1 - p_2 > p_0,$$

where the hypothesized difference, p_0, is typically but not necessarily zero.

[14] Analogous to the single proportion case, $q_1 = 1 - p_1$ and $q_2 = 1 - p_2$ represent the true proportions of failures.

computed using[15]

$$z^* = \frac{\hat{p}_1 - \hat{p}_2}{\sqrt{\bar{p}\,\bar{q}\left(\dfrac{1}{n_1} + \dfrac{1}{n_2}\right)}},$$

where

$$\bar{p} = \frac{x_1 + x_2}{n_1 + n_2},$$

$\bar{q} = 1 - \bar{p}$, and z^* has an approximate standard normal distribution. The corresponding p-value is then given by

$$\text{p-value} = \begin{cases} P(Z \geq |z^*|) & \text{for a one-tailed test} \\ 2\,P(Z \geq |z^*|) & \text{for a two-tailed test} \end{cases}$$

Example

To test the previously stated hypotheses,

$$H_0 : p_{\text{S}} \leq p_{\text{J}} \qquad \text{vs.} \qquad H_1 : p_{\text{S}} > p_{\text{J}} \quad \text{(Claim)},$$

basic code can be used to first calculate $\hat{p}_{\text{S}}$, $\hat{p}_{\text{J}}$, $\bar{p}$ and $\bar{q}$, and then compute $z^* = 1.658$, p-value $= 0.0487$. So, for any $\alpha \geq 0.05$, there is sufficient evidence to support the claim that the proportion of brown bears is higher in the Sitka area than in the Juneau area.

11.3.4 A Chi-Square Test

The strategy is analogous to that of the single proportion case, and the test is equivalent to the previous two-sample z-test. First use the *observed frequencies* and the null hypothesis to find the *expected frequencies*. Let n_1 and n_2 denote the number of trials for the two samples, let $n = \sum_{j=1}^{2} n_j$, and organize the observed frequencies, o_{ij}, in a table as follows

	1	2	Totals
Success	o_{11}	o_{12}	r_1
Failure	o_{21}	o_{22}	r_2
Totals	c_1	c_2	n

Then find the *expected frequencies*, e_{ij}, for each corresponding cell as follows. Under the null hypothesis $p_1 = p_2$ (and, consequently, $q_1 = q_2$ as well),[16]

---

[15]For hypotheses as described in Footnote 13, if $p_0 \neq 0$ the numerator is replaced by $(\hat{p}_1 - \hat{p}_2) - p_0$.

[16]Note that while the null hypothesis is often written in one of three forms, that is, $p_1 \geq p_2$, $p_1 = p_2$, or $p_1 \leq p_2$, the form $p_1 = p_2$ is used in developing the test.

therefore for each $i = 1, 2$ and $j = 1, 2$,

$$\frac{e_{ij}}{c_j} = k_i \quad \text{(a constant for each } i\text{)}.$$

So $e_{ij} = k_i \, c_j$, and the row totals are given by

$$r_i = \sum_{j=1}^{2} k_i \, c_j = k_i \, n.$$

Consequently,

$$\frac{r_i}{n} = \frac{e_{ij}}{c_j},$$

and $e_{ij} = r_i \, c_j / n$ provides the expected frequencies to compute the test statistic

$$\chi^{2*} = \sum_{i=1}^{2} \sum_{j=1}^{2} \frac{(o_{ij} - e_{ij})^2}{e_{ij}}.$$

If the samples are independent and if all the expected frequencies satisfy $e_{ij} \geq 5$, then $\chi^{2*} \sim \chi^2(1)$,[17] and the p-value for this test is obtained as previously described. For two-tailed tests use

$$\text{p-value} = P(\chi^2 \geq \chi^{2*});$$

and for a one-tailed test, first compute

$$z^* = \sqrt{\chi^{2*}},$$

then use the standard normal distribution to compute

$$\text{p-value} = P(Z \geq z^*)$$

irrespective of whether the test is left- or right-tailed.

The function `prop.test` can be used here too, the general usage definition for this scenario being of the form

```
prop.test(x, n, p,
          alternative, conf.level, correct = TRUE)
```

Here the number of observed successes for the two samples are entered into the function using an argument assignment of the form x = c(x_1, x_2) and, similarly, n = c(n_1, n_2) provides the number of trials performed on each

---

[17]The contingency table of observed frequencies for such cases has 2 rows and 2 columns. For a computed statistic to remain constant, observations in one row and one column are fixed and the remaining observation is allowed to vary. So, the chi-square distribution associated with the statistic has $\nu = 1$ degree of freedom.

population.[18] Since the default setting for p is 0 for a two proportions case, a hypothesized difference in the proportions typically need not be assigned to p. The remaining arguments are treated in the same manner as for the single proportion case.[19]

Example

Results equivalent to those produced in the previous example are obtained using code of the form

```
> prop.test(x = bears["Brown",], n = bears["Total",],
+     alternative = "greater", correct = FALSE)
```

to get $\chi^{2*} = 2.7489$, $\nu = 1$ and p-value $= 0.04866$. Once again, for any $\alpha \geq 0.05$, there is sufficient evidence to support the claim.

A Comment about the Confidence Interval Produced: Since the test is right-tailed the default 95% confidence interval produced by prop.test, rounded to 3 decimal places, is

$$0.003 < p_1 - p_2 \leq 1.$$

The right-bound is set at 1 since a difference between two proportions cannot be greater than 1. Since the hypothesized value, $p_0 = p_1 - p_2 = 0$, just barely does not lie inside the confidence interval, the null hypothesis is rejected at $\alpha = 0.05$.

11.3.5   Fisher's Exact Test

There may be occasions where the data in a 2×2 contingency table do not truly represent samples associated with two independent and binomially distributed random variables, the expectation for the previously described equivalent methods. For example, some experiments necessitate the use of samples of predetermined sizes from a group of N available subjects which are randomly assigned to two treatments; see, for example, [2, p. 95] or [27, pp. 188–191].

Recall the previous 2×2 contingency table with some added notation for clarity in the following discussion.

---

[18]Alternatively, this function can be used by assigning x a 2×2 matrix in which the first column contains the number of successes and the second column the number of failures. In this case n is left out of the function call. The remaining arguments are assigned values as desired.

[19]According to the details provided for the function prop.test, the default setting for the continuity correction argument is correct = TRUE and under this setting (as applicable to a two proportions test) the correction $\varepsilon = 1/(2n)$ is applied only if $\varepsilon \leq |\hat{p}_1 - \hat{p}_2|$.

	Treatment 1	Treatment 2	Totals
Outcome A	$x = o_{11}$	$k - x = o_{12}$	$k = r_1$
Outcome B	o_{21}	o_{22}	$N - k = r_2$
Totals	$m = c_1$	$N - m = c_2$	$N = c_1 + c_2$

The rows in this table identify the outcomes to which the subjects of the two treatments find themselves associated with. The row 1 sum $k \le N$ represents the number of subjects who end up associated with Outcome A, x of whom were assigned to Treatment 1 and $k - x$ of whom were assigned to Treatment 2. The remainder are associated with Outcome B. Observe that if m and k are predetermined, then the row and column sums are not random numbers.

For small populations or collections (in particular, when $N < 20$) it is recommended that *Fisher's exact test* be used for tests of two proportions. This is even if the row sums (r_1 and r_2) and column sums (c_1 and c_2) of the associated 2 × 2 contingency are not *both* predetermined, see [123, p. 371].[20]

For Fisher's exact test the associated random variable has a hypergeometric distribution with $X \sim HYP(k, m, N)$. The random variable X represents the number of Treatment 1 subjects (out of m) that are selected from the N objects through k trials, and that fall under Outcome A. Recall that for $X \sim HYP(k, m, N)$ non-zero probabilities occur for

$$\max(0, k - N + m) \le x \le \min(k, m)$$

and the number of trials may range over $k = 1, 2, \ldots, N$. The test statistic, $x^* = o_{11}$, for Fisher's exact test is the observed frequency of Treatment 1 subjects associated with Outcome A and, using

$$P(X = x) = \binom{m}{x} \binom{N - m}{k - x} \bigg/ \binom{N}{k},$$

the p-value is computed as follows: For a left-tailed alternative use p-value $= P(X \le x^*)$, and for a right-tailed alternative use p-value $= P(X \ge x^*)$. For a two-tailed alternative use

$$\text{p-value } = 2 \min \{P(X \le x^*), P(X \ge x^*)\},$$

and if this results in a number greater than 1, then set p-value $= 1$.

The R function for this test, along with arguments of interest, is

```
fisher.test(x, y = NULL, alternative = "two.sided",
    conf.int = TRUE, conf.level = 0.95)
```

---

[20]This test is discussed in some depth in [27, pp. 189–191] where it is indicated that if either the row sum, or the column sum, or both are random numbers (not predetermined) then Fisher's test as described here is no longer an exact test. It is however still a valid, but conservative, test that is commonly used for small sample tests associated with 2 × 2 contingengy tables. See, also, [2, pp. 90–96].

where x may be assigned a 2×2 matrix containing the above contingency table. If x is assigned a matrix, then y is ignored (not needed). Alternatively, these data can be viewed as ordered pairs of nominal bivariate data (the corresponding random variables being Category and Class). In this case x is assigned a vector listing the (column) categories to which the objects belong and y is assigned a vector listing the corresponding (row) class to which each object is associated with.

Fisher's Test and the Odds Ratio

The function `fisher.test` reports the same p-value, but views the null (and alternative) hypotheses in terms of the ratio of the odds of obtaining Outcome A in Treatment 1 to the odds of obtaining Outcome A in Treatment 2. That is, the odds ratio, $\theta = (p_1/q_1)/(p_2/q_2)$, where

$$p_1 = P(\text{Outcome } A \mid \text{Treatment 1}), \quad q_1 = P(\text{Outcome } B \mid \text{Treatment 1}),$$
$$p_2 = P(\text{Outcome } A \mid \text{Treatment 2}), \quad q_2 = P(\text{Outcome } B \mid \text{Treatment 2}).$$

So, the hypotheses appear as, for example,[21]

$$H_0 : \theta = 1 \quad \text{vs.} \quad H_1 : \theta \neq 1.$$

The *unconditional maximum likelihood estimate of* θ, or *sample odds ratio* is given by,

$$\hat{\theta} = \frac{o_{11}o_{22}}{o_{12}o_{21}}.$$

However, it is important to note that the interval and point estimates of the odds ratio θ reported in the output produced by `fisher.test` are obtained using the *conditional maximum likelihood estimate* (rather than the sample odds ratio), that is, that value of θ which maximizes the conditional maximum likelihood function

$$L(x^* \mid k, m, N : \theta) = \binom{m}{x^*}\binom{N-m}{k-x^*}\theta^{x^*} \Bigg/ \sum_{z=a}^{b} \binom{m}{z}\binom{N-m}{k-z}\theta^z,$$

where $x^* = o_{11}$, $a = \max(0, k - N + m)$ and $b = \min(k, m)$. See the Figure 11.5 for a graphical illustration of this.

---

[21]The equivalence of the hypotheses
$$H_0 : \theta = 1 \quad \text{vs.} \quad H_1 : \theta \neq 1$$
and
$$H_0 : p_1 = p_2 \quad \text{vs.} \quad H_1 : p_1 \neq p_2$$
can be verified as follows. Using the definition
$$\theta = \frac{p_1/q_1}{p_2/q_2}$$
under the null hypothesis and subject to the conditions that $0 < p_1 < 1$ and $0 < p_2 < 1$,
$$\theta = 1 \quad \Longleftrightarrow \quad \frac{p_1}{1-p_1} = \frac{p_2}{1-p_2} \quad \Longleftrightarrow \quad p_1 = p_2.$$

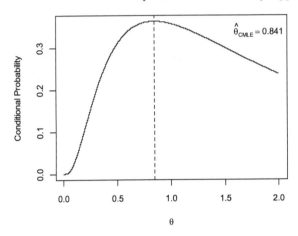

FIGURE 11.5

Plot of the maximum likelihood function used to obtain the conditional estimate of the odds ratio for the first of the following examples illustrating Fisher's test. For this likelihood function the maximizing odds ratio θ is estimated when $x^* = 5$, $k = 8$, $m = 13$ and $N = 20$. See the script file for further examples and the code used.

The use of the conditional maximum likelihood estimate is recommended if the contingency table does not represent or even approximate binomial samples. This is definitely the case for small sample scenarios described in the preliminary paragraph of this section. Further details on this matter can be found in, for example, [2, p. 606] and [64].

There are three possible scenarios in which an application of Fisher's exact test might find use. The first example illustrates a scenario in which Fisher's test functions as an exact test.

Example: Column and Row Sums Are Both Predetermined

Consider the case in which the manager of two coffee shops is asked to distribute 20 new baristas among the two shops. Shop A has 13 openings and Shop B has 7, and 8 of the new employees are experienced whereas 12 are not. The results of the assignments are shown below, and also placed in a matrix with margins named `newHires`.

	Shop A	Shop B	Totals
Experienced	5	3	8
Not Experienced	8	4	12
Totals	13	7	20

One might ask if there is evidence to suggest that the manager's assignment method favored one coffee shop over the other with respect to experienced hires.

Since the number of openings at each shop are known and the number of experienced versus inexperienced new hires are known, Fisher's test, as applied to this case, is an exact test.

Let p_A and p_B represent the true proportions of experienced hires that may be assigned to Shops A and B, respectively, if a truly random process is followed, then the hypotheses to be tested are

$$H_0 : p_A = p_B \quad \text{vs.} \quad H_1 : p_A \neq p_B.$$

The test statistic, $x^* = 5$, belongs to $X \sim HYP(8, 13, 20)$ and, being a two-tailed test, the p-value is computed using

$$\text{p-value} = 2 \min \{P(X \leq x^*), P(X \geq x^*)\}.$$

The hypergeometric cumulative distribution function and basic code can be used directly to get $x^* = 5$, $k = 8$, $m = 13$, $N = 20$, $\hat{\theta} = 0.8333$ and p-value $= 1$ (see the script file for the code).

Alternatively, the function `fisher.test` with the default two-sided alternative can be run (see, also, Figure 11.5).

```
> fisher.test(x = newHires[-3,-3])
```

Here too $x^* = 5$, $k = 8$, $m = 13$, $N = 20$ and p-value $= 1$. But, in this case the odds ratio is $\hat{\theta} = 0.8410$. So, there is certainly not enough evidence to suggest that the manager's assignment method favored one coffee shop over the other with respect to experienced hires.

Note that the assignment `x = newHires[-3,-3]` excludes the margins of `newHires` from being passed into `x`.

Example: Only Column (or Row) Sums Are Predetermined

Consider the case in which a convenience sample of 20 bears, 11 black and 9 brown, on a Southeast Alaska island undergo a psychiatric evaluation of their respective temperaments. The data containing the resulting classifications are shown below, and placed in a matrix with margins named `moodyBears`.

	Black	Brown	Totals
Temperamental	5	3	8
Mellow	3	6	9
Totals	8	9	17

Based on their experiences, people who live on the island believe that black bears are more temperamental than brown bears. Do the data support this belief?

Let p_{Black} and p_{Brown} represent the true proportions of black and brown bears, respectively, on this island that are temperamental. Then, the hypotheses to be tested are

$$H_0 : p_{\text{Black}} \leq p_{\text{Brown}} \quad \text{vs.} \quad H_1 : p_{\text{Black}} > p_{\text{Brown}}.$$

The test statistic, $x^* = 5$, belongs to $X \sim HYP(8, 8, 17)$, and the right-tailed alternative yields p-value $= 0.2380$ – the code is similar to that used for the previous example.

To use the `fisher.test` function, run

```
> fisher.test(x = moodyBears[-3,-3], alternative = "greater")
```

to get $\hat{\theta} = 3.091475$ and p-value $= 0.238$. So, at least based on the sample from the island, there is insufficient evidence to support the islanders' belief about the temperament of black and brown bears on their island.

Example: Neither Column nor Row Sums Are Predetermined

Consider the case in which a group of 30 mathematics and other college students unknowingly enrolled in a project-based elementary statistics course; the format and delivery mode of the course was not advertised. All 30 students completed the course, and at the end of the course they were asked to identify themselves as a mathematics major (or not) and state whether they enjoyed the course (or not). The data are shown below, and placed in the matrix `enjoyStats` with margins.

	Mathematics	Other	Totals
Enjoyed	11	13	24
Did not enjoy	3	3	6
Totals	14	16	30

Suppose the general belief is that, on average, mathematics majors tend *not* to enjoy project-based statistics courses as much as non-mathematics majors.

To use the data to test this belief, let p_{M} and p_{O} represent the true proportions of mathematics majors and all other majors, respectively, who enjoy project-based elementary statistics courses. The hypotheses to be tested are

$$H_0 : p_{\text{M}} \geq p_{\text{O}} \quad \text{vs.} \quad H_1 : p_{\text{M}} < p_{\text{O}},$$

the test statistic and relevant parameters are $x^* = 11$, $k = 24$, $m = 14$, $N = 30$, and running

```
> fisher.test(x = enjoyStats[-3,-3], alternative = "less")
```

gives $\hat{\theta} = 0.8509$ and p-value $= 0.6046$. This indicates, based on the data, that there is insufficient evidence to support the belief that mathematics majors tend not to enjoy project-based statistics courses as much as non-mathematics majors.

11.3.6 Which Approach Should Be Used?

In scenarios similar to the previous three examples, where $N \leq 20$ or the expected frequencies are not all at least 5, Fisher's (exact) test should be employed with the understanding that this is a conservative test.

It is known that if m/N approaches a unique positive constant p where $0 \leq p \leq 1$ as both m and N grow large the random variable $X \sim HYP(k, m, N)$ can be approximated by the binomial distribution $BIN(k, p)$ (see, for example, [7, p. 97]). Additionally, as mentioned previously, and particularly if the row and column sums of the contingency table are not both predetermined, Fisher's test is more conservative than the z- and chi-squared tests. For this reason one of the two (equivalent) approximate methods should be preferred over Fisher's test (see, for example, [27, pp. 188–191]).

11.4 Additional Notes

It is helpful to have a feel for how the normal approximation of binomial distributions works; an illustration of this follows. Also, a question that frequently comes up is the matter of when it is appropriate to use a one-sided hypothesis test, a choice that is discouraged by some. Suggestions about this matter are provided in the last sub-section.

11.4.1 Normal Approximations of Binomial Distributions

Let $X \sim BIN(n, p)$ and for any observed $X = x$ let $\hat{p} = x/n$, then the sampling distribution of $\hat{p}$ has a mean of $\mu_{\hat{p}} = p$ and variance $\sigma_{\hat{p}}^2 = p(1 - p)/n$ and, by the Central Limit Theorem, as n increases the distribution of $\hat{p}$ approaches that of a normal distribution. That is,

$$\hat{z} = \left(\hat{p} - \mu_{\hat{p}}\right) / \sigma_{\hat{p}} \rightarrow N(0, 1)$$

asymptotically, and the rule of thumb provided is that this approximation may be assumed acceptably close if both $np \geq 5$ and $n(1 - p) \geq 5$ hold true; this is particularly the case if p is close to 0.5. See the script file for code to illustrate this graphically.

11.4.2   One- versus Two-Sided Hypothesis Tests

A concise and practical discussion on this topic appears in [112] and a longer but equally enlightening philosophical discussion extending to multiple pairwise tests appears in [133]. Here, the matter of one- versus two-sided tests is summarized from two points of view: the quantitative consequences of choosing one over the other, and the underlying theoretical justification for a particular choice summarized from [112] (see also [76] and [103]).

Consider the quantitative point of view first and, for simplicity, use the z-test (normal approximation approach) to compare two population proportions p_1 and p_2. Whether the test to be conducted is one-sided or two-sided, the test statistic is

$$z^* = \frac{\hat{p}_1 - \hat{p}_2}{\sqrt{\bar{p}\bar{q}\left(\dfrac{1}{n_1} + \dfrac{1}{n_2}\right)}}$$

with

$$\bar{p} = \frac{x_1 + x_2}{n_1 + n_2}.$$

Without loss of generality, suppose $z^* > 0$, and let z_α and $z_{\alpha/2}$ be solutions of $P(Z \geq z_\alpha) = \alpha$ and $P(Z \geq z_{\alpha/2}) = \alpha/2$, respectively. Now consider the alternatives

$$H_1 : p_1 - p_2 > 0 \quad \text{and} \quad H_1 : p_1 - p_2 \neq 0.$$

Regardless of which alternative is involved, there are three possible scenarios,

$$z^* < z_\alpha < z_{\alpha/2}, \quad \text{or} \quad z_\alpha \leq z^* < z_{\alpha/2}, \quad \text{or} \quad z_\alpha < z_{\alpha/2} \leq z^*.$$

For the first, the null hypothesis is not rejected whether or not the alternative is one- or two-sided; and for the third, the null hypothesis is rejected for both the one- and two-sided alternatives. For the middle (second) scenario, the null hypothesis is rejected for the one-sided alternative but not for the two-sided alternative. So, a one-sided test is more likely to result in the rejection of a false null hypothesis than a two-sided test: a one-sided has more power than a two-sided test.

There are two reasons that can be given for choosing (and justifying) a one-sided test over a two-sided test. First, if there is a strong argument (based on underlying theory and/or anecdotal evidence) in favor of using a one-sided alternative (for example, $p_1 - p_2 > 0$); *and* second, in the unexpected event that the null hypothesis is actually true (that is, $p_1 - p_2 \leq 0$), there is a strong argument as to why an inequality in the direction opposite to the alternative (that is, $p_1 - p_2 < 0$) is not of concern. If either one of these two reasons cannot be supported through sound justifications, a two-sided should be used.

Another suggestion provided, for example in [2, p. 92] and echoed in [123, p. 370], to guard against criticism is as follows. If a one-sided alternative is suspected, then to perform a one-sided test at an α level of significance, perform a two-sided test at the 2α level of significance.

12

Tests for More than Two Proportions

For this chapter the focus is on testing hypotheses involving three or more population proportions based on observed frequencies associated with categorical variables. It is worth noting that some of these methods apply equally well to scenarios with two proportions. As for the relevant probability distributions for this chapter, the chi-square distribution is the only player; see Section 11.1.4 for material on this distribution. As for past chapters, code for all examples and a bit more are contained in the file `Chapter12Script.R`.

12.1 Equality of Three or More Proportions

Of interest here is determining if there is evidence to reject an assumption of equality of $J \geq 3$ population proportions. In particular, if $p_1, p_2, \ldots, p_J$ represent the true population proportions across the J classes, then the hypotheses to be tested are

$$H_0 : p_1 = p_2 = \cdots = p_J \quad \text{vs.} \quad H_1 : \text{The proportions are not all equal}$$

Note that the alternative hypothesis includes a wide range of possibilities. So, if the null hypothesis is rejected all that can be said is that the statement $p_1 = p_2 = \cdots = p_J$ is (most likely) false. Section 12.2 addresses methods that can be used to determine the nature of differences that may exist between pairs of the J population proportions, and Section 12.3 addresses what are called general linear contrasts of the J proportions. Three procedures for testing such hypotheses are presented here, the first of which is commonly encountered in elementary statistics texts.

12.1.1 Pearson's Homogeneity of Proportions Test

The chi-square *homogeneity of proportions test* is the commonly encountered method; the two-sided test for the previous two proportions case is a special case of this method. Data for such tests can be viewed in the form of a 2-dimensional contingency table with two rows, the first row representing observed frequencies for successes, and the second row representing frequencies of failures.

	1	2	$\cdots$	J	Totals
Success	o_{11}	o_{12}	$\cdots$	o_{1J}	r_1
Failure	o_{21}	o_{22}	$\cdots$	o_{2J}	r_2
Totals	c_1	c_2	$\cdots$	c_J	n

The columns represent the J classes of a variable or J populations of interest. Analogous to the two proportions case, for $i = 1, 2$ and $j = 1, 2, \ldots, J$, the corresponding expected frequencies are obtained using $e_{ij} = r_i\, c_j / n$ and the test statistic is

$$\chi^{2*} = \sum_{i=1}^{2} \sum_{j=1}^{J} \frac{(o_{ij} - e_{ij})^2}{e_{ij}}.$$

If every expected frequency satisfies $e_{ij} \geq 5$, then $\chi^{2*} \sim \chi^2(J - 1)$ and the p-value is obtained using[1]

$$\text{p-value } = P(\chi^2 \geq \chi^{2*}).$$

While the `prop.test` function can be used to perform such tests,[2] this is an opportunity to introduce another function, `chisq.test`. Code to use this function for the present test is

```
chisq.test(x)
```

where x is assigned the $2 \times J$ matrix of observed frequencies. A correction for continuity is never used for such applications of `chisq.test` and `prop.test`, and all remaining arguments (not shown) are left in their default settings.

Example

One hundred and seventy five students were randomly assigned to one of two elementary statistics courses, one traditional and the other project-based, both cover the same content. At the end of the course all students sat for a common comprehensive final examination under identical conditions. Data on these students are contained in the data frame `elemStatStudy`. The fields of study of students enrolled in these two courses included the mathematical sciences, natural sciences, and the social sciences.

---

[1] The contingency table of observed frequencies for such cases has 2 rows and J columns. For a computed statistic to remain constant, observations in one row and one column are fixed and the rest are allowed to vary. So, the chi-square distribution associated with the statistic has $\nu = J - 1$ degrees of freedom.

[2] The transpose of the contingency table of observed frequencies can be passed into the function as follows

```
prop.test(x = t(o))
```

Here, `t(o)` is the transpose of the matrix o, and contains the number of successes in the first column and the number of failures in the second. All remaining arguments are left in their default settings. See the script file for an example.

Let p_M, p_N and p_S represent the (unknown) expected proportions of students in the mathematical sciences, natural sciences and social sciences, respectively, who score 85% or higher on the final examination after taking the project-based course.

Suppose the intent here is to determine if students from the three broad subject areas respond differently to the project-based course, then the hypotheses to be tested are

$$H_0 : p_M = p_N = p_S \quad \text{vs.} \quad H_1 : \text{The proportions are not all equal.}$$

To test these hypotheses, first place the observed frequencies of interest in a contingency table, o, and then run

```
> chisq.test(x = o)
```

to get $\chi^{2*} = 3.0059$, $\nu = 2$ and p-value $= 0.2225$. So, at any typical level of significance, there is insufficient evidence to conclude that students from the three areas respond differently to the project-based course.

What If the Expected Frequencies Condition Is Violated?

Both the chisq.test and prop.test functions issue a warning if any of the cells have an expected frequency of less than 5. A general (and conservative) rule-of-thumb is that the percentage of cells with an expected frequency of less than 5 should be below 20%, and none of these cells should have an expected frequency of less than 1 (see, for example, [27, pp. 201–202] or [123, pp. 360–361]).[3]

If the expected frequency for one (or more) cells is less than 5, and a larger sample is not possible or helpful, there are a couple of options. One suggestion (see, for example, [12, Sec. 11.2]) is to combine classes having low observed frequencies with other classes for which the class characteristics might be considered closely allied (or similar). Another option would be to use Cohen's small sample procedure described in Section 12.1.3.

12.1.2   Marascuilo's Large Sample Procedure

This was presented by Marascuillo in [87, Ex. 2] as a large-sample alternative to the previously described Pearson's chi-square homogeneity of proportions procedure. Marascuilo describes this procedure as a chi-square analog of Sheffe's F-test procedure for testing contrasts of means (see, for example, [80, Sec. 17.6] or [118, Ch. 3]).

The hypotheses for this procedure are the same as for Pearson's procedure.

---

[3]Quick sample code to determine this percentage, if a warning is issued, is

```
E <- chisq.test(x = o)$expected; round(sum(E < 5)/prod(dim(E))*100)
```

Where this procedure differs from Pearson's procedure is in how the test statistic is computed.

For $j = 1, 2, \ldots, J$, let $\hat{p}_j$ represent the unbiased and independent large sample (asymptotically normally distributed) estimate of p_j, and

$$s_{\hat{p}_j}^2 = \frac{\hat{p}_j \hat{q}_j}{n_j}$$

the sample estimate of $\sigma_{\hat{p}_j}^2$. Now, with $a_j = 1/s_{\hat{p}_j}^2$, define the estimate for the (unknown) proportion for which $p_1 = p_2 = \cdots = p_J = p_0$ by

$$\hat{p}_0 = \sum_{j=1}^{J} a_j \hat{p}_j \Big/ \sum_{j=1}^{J} a_j.$$

Then the test statistic is computed using

$$\chi^{2*} = \sum_{j=1}^{J} \frac{(\hat{p}_j - \hat{p}_0)^2}{s_{\hat{p}_j}^2},$$

and, for large samples, $\chi^{2*} \sim \chi^2(J-1)$ allows for computing

$$\text{p-value} = P(\chi^2 \geq \chi^{2*}).$$

For convenience, these computations are packaged in the function

```
marascuiloProp.tests(x, print.what = "both", conf = 0.95)
```

As for the function chisq.test, the argument x is assigned a contingency table in which the first row contains the number of successes for each class, and the second row the number of failures.

The argument print.what is used to indicate what should be outputted. In the default setting results of the omnibus test described above as well as intervals of simultaneous pairwise differences[4] at a desired confidence level are outputted (the default setting is conf = 0.95). To output only the results of the omnibus test, use the argument assignment print.what = "omnibus" and use print.what = "pairwise" to output only details on pairwise differences. For the curious, code for this function is contained at the end of the script file for this chapter.

Example

This method can be applied to the example described previously for Pearson's procedure using the marascuiloProp.tests function by running

```
> marascuiloProp.tests(x = o, print.what = "omnibus")
```

to get $\chi^{2*} = 3.286$, $\nu = 2$ and p-value $= 0.1934$. As should be expected, the result does not contradict the result from Pearson's procedure. The test statistic from this procedure, however, is larger than that obtained using Pearson's procedure.

---

[4]Details on these are given in in Section 12.2.1.

12.1.3 Cohen's Small Sample Procedure

This procedure was proposed by Cohen as an alternative to Marascuilo's previously mentioned procedure, see [25] or [26, Ch. 6 and Sec. 12.6]. It is worth noting that this procedure is independent of the large sample requirement for Marascuilo's method, so it can be used with small and large samples. Cohen also states that this procedure can even be used if some of the observed frequencies are zero. Here are the computational steps.

Let n_j, $\hat{p}_j$, and p_j be as defined for Marascuilo's procedure. Using *Fisher's arcsine transformation of proportions*, define

$$\hat{\phi}_j = \begin{cases} 2\arcsin\sqrt{1/(4n_j)} & \text{if } \hat{p}_j = 0, \\ 2\arcsin\sqrt{\hat{p}_j} & \text{if } 0 < \hat{p}_j < 1, \\ \pi - 2\arcsin\sqrt{1/(4n_j)} & \text{if } \hat{p}_j = 1. \end{cases}$$

Then each $\hat{\phi}_j$ is approximately normally distributed and an unbiased estimate of $\sigma_{\hat{\phi}_j}^2$ is given by $s_{\hat{\phi}_j}^2 = 1/n_j$. The equivalent null hypothesis for this procedure is

$$H_0 : \phi_1 = \phi_2 = \cdots = \phi_J = \phi_0$$

with

$$\hat{\phi}_0 = \sum_{j=1}^{J} n_j \hat{\phi}_j / n,$$

and the test statistic is

$$\begin{aligned} \chi^{2*} &= \sum_{j=1}^{J} \frac{\left(\hat{\phi}_j - \hat{\phi}_0\right)^2}{s_{\hat{\phi}_j}^2} \\ &= \sum_{j=1}^{J} n_j \left(\hat{\phi}_j - \hat{\phi}_0\right)^2. \end{aligned}$$

Again, $\chi^{2*} \sim \chi^2(J-1)$ and p-value $= P(\chi^2 \geq \chi^{2*})$ and these computations are packaged in the function

```
cohenProp.tests(x, print.what = "omnibus", conf = 0.95)
```

The arguments and default settings are as for the `marascuiloProp.tests` function. Code for this function is also contained at the end of the script file for this chapter.

Example

Applying this procedure to the previous example is a matter of running

```
> cohenProp.tests(x = o, print.what = "omnibus")
```

to get $\chi^{2*} = 3.1442$, $\nu = 2$ and p-value $= 0.2076$. Once again, the result does not contradict that obtained using Pearson's procedure. In this case the test statistic lies between those of the previous two procedures.

12.2   Simultaneous Pairwise Comparisons

If the null hypothesis for one of the previously described omnibus tests is rejected, then it may be of interest to determine the nature of the differences by performing simultaneous pairwise comparisons of proportions. Since there are J classes, there are $\binom{J}{2}$ distinct pairs to be compared and, for each pair $j, k = 1, 2, \ldots, J$ with $j \neq k$, the relevant hypotheses to be tested are

$$H_0 : p_j = p_k \quad \text{vs.} \quad H_1 : p_j \neq p_k.$$

Pearson's test from Section 12.1.1 can be applied at the same significance level for this purpose (see, for example, [27, p. 220] for an analogous extension). Alternatively, special cases of both Marascuilo's and Cohen's procedures can be used for this purpose, see [87, Ex. 2] and [25].

See also Section 15.7 for further ideas on interpreting the nature of differences that may be found using any one of the procedures discussed here.

12.2.1   Marascuilo's Large Sample Procedure

Denote the sample proportions of success and failure by $\hat{p}_j, \hat{p}_k$ and $\hat{q}_j, \hat{q}_k$, respectively. Then, assuming the sample sizes are sufficiently large, at the α level of significance the null hypothesis for the pair (p_j, p_k) is rejected in favor of the alternative if zero does not lie in the interval

$$(\hat{p}_j - \hat{p}_k) - \sqrt{\chi^2_\alpha} \sqrt{\frac{\hat{p}_j \hat{q}_j}{n_j} + \frac{\hat{p}_k \hat{q}_k}{n_k}}$$

$$< p_j - p_k <$$

$$(\hat{p}_j - \hat{p}_k) + \sqrt{\chi^2_\alpha} \sqrt{\frac{\hat{p}_j \hat{q}_j}{n_j} + \frac{\hat{p}_k \hat{q}_k}{n_k}},$$

where χ^2_α satisfies $P(\chi^2 \geq \chi^2_\alpha) = \alpha$ using the chi-square distribution with $\nu = J - 1$ degrees of freedom.

Note that this is equivalent to using

$$\chi^{2*} = \frac{(\hat{p}_j - \hat{p}_k)^2}{\hat{p}_j \hat{q}_j / n_j + \hat{p}_k \hat{q}_k / n_k}$$

and the chi-square distribution having $\nu = J - 1$ degrees of freedom to compute p-value $= P(\chi^2 \geq \chi^{2*})$, and then comparing the result against α.

Example

The computations, at the 90% confidence level, can be performed on the data used in the previous examples using

```
> marascuiloProp.tests(x = o, print.what = "pairwise",
       conf = 0.90)
```

It will be observed that the results do not contradict that of the previously performed Marascuilo's omnibus test.

12.2.2 Cohen's Small Sample Procedure

Using the previously defined arcsine transformation for Cohen's procedure at the α level of significance, the null hypothesis is rejected in favor of the alternative if zero does not lie in the interval

$$(\hat{\phi}_j - \hat{\phi}_k) - \sqrt{\chi_\alpha^2}\sqrt{\frac{1}{n_j} + \frac{1}{n_k}}$$

$$< \phi_j - \phi_k <$$

$$(\hat{\phi}_j - \hat{\phi}_k) + \sqrt{\chi_\alpha^2}\sqrt{\frac{1}{n_j} + \frac{1}{n_k}},$$

where χ_α^2 satisfies $P(\chi^2 \geq \chi_\alpha^2) = \alpha$ using $\chi^2(J-1)$. Equivalently, compute

$$\chi^{2*} = \frac{(\hat{\phi}_j - \hat{\phi}_k)^2}{1/n_j + 1/n_k}$$

and then use $\chi^2(J-1)$ to compute and compare p-value $= P(\chi^2 \geq \chi^{2*})$ against α.

Example

Using the same data as for the previous examples, and at the 90% confidence level, the computations are performed using

```
> cohenProp.tests(x = o, print.what = "pairwise",
+      conf = 0.90)
```

The results do not contradict the the previously performed Cohen's test.

12.3 Linear Contrasts of Proportions

Let $p_1, p_2, \ldots, p_J$ represent the true proportions of successes across $J \geq 3$ classes from which samples of sizes $n_1, n_2, \ldots, n_J$ are collected. Define

$$\psi = c_1 p_1 + c_2 p_2 + \cdots + c_J p_J$$

where $c_1, c_2, \ldots, c_J$ are real numbers that satisfy $c_1 + c_2 + \cdots + c_J = 0$. Then ψ is called a *linear contrast* of the proportions $p_1, p_2, \ldots, p_J$ and the associated hypotheses take the form

$$H_0 : \psi = \psi_0 \quad \text{vs.} \quad H_1 : \psi \neq \psi_0,$$

where ψ_0 represents the hypothesized value of ψ. Typically, it is the case that $\psi_0 = 0$. Note that the pairwise differences tested in the previous section are contrasts too; indeed, for $\psi = p_j - p_k$, the coefficients on the right side add to zero. These are referred to as as *simple contrasts*.

12.3.1   Marascuilo's Large Sample Approach

This approach is described in [87, Ex. 2]. Let ψ be a linear contrast of the proportions $p_1, p_2, \ldots, p_J$. Suppose $\hat{p}_1, \hat{p}_2, \ldots, \hat{p}_J$ are the unbiased, independent large sample (asymptotically normally distributed) estimates of $p_1, p_2, \ldots, p_J$, then

$$\hat{\psi} = c_1 \hat{p}_1 + c_2 \hat{p}_2 + \cdots + c_J \hat{p}_J$$

is the same for ψ. Now, denote the sample estimate of $\sigma_{\hat{\psi}}^2$ by $s_{\hat{\psi}}^2$. Since the $\hat{p}_1, \hat{p}_2, \ldots, \hat{p}_J$ are independent,

$$s_{\hat{\psi}}^2 = c_1^2 \hat{p}_1 \hat{q}_1 / n_1 + c_2^2 \hat{p}_2 \hat{q}_2 / n_2 + \cdots + c_J^2 \hat{p}_J \hat{q}_J / n_J,$$

and a $(1 - \alpha) \times 100\%$ confidence interval of ψ is given by

$$\hat{\psi} - \sqrt{\chi_\alpha^2}\, s_{\hat{\psi}} < \psi < \hat{\psi} + \sqrt{\chi_\alpha^2}\, s_{\hat{\psi}},$$

where, using the chi-square distribution with $\nu = J - 1$ degrees of freedom, χ_α^2 satisfies $P(\chi^2 \geq \chi_\alpha^2) = \alpha$.

Now, let ψ_0 denote a hypothesized value of ψ, then the hypotheses associated with the test corresponding to the above interval are as stated in the introductory comments of this section. So, for the chosen level of significance, the null hypothesis is rejected if ψ_0 does not lie in the interval for ψ. The statistic corresponding to this test is

$$\chi^{2*} = \frac{\left(\hat{\psi} - \psi_0\right)^2}{s_{\hat{\psi}}^2} \sim \chi^2(J - 1)$$

where p-value $= P(\chi^2 \geq \chi^{2*})$ is computed in the usual manner.

Example

The new manager of the Wet Weather Warehouse, which supplies much of Juneau's wet-weather gear, is looking to cut down on the variety of brand-names sold at each of WWW's five smaller outlets in the Juneau area. It has been observed over time that the average monthly revenue for stores 1 and 2 approximately equals the average monthly revenue for the other three stores. The manager has decided to either turn stores 1 and 2 into single brand-name stores, or do the same to stores 3, 4 and 5. Each store has a loyal customer base and, total revenue being the bottom line, the task then is to determine how the customers might react. Random samples of customers from each store were asked for their reaction to the proposal, the response options being "Unhappy" or "Do not care." The data are contained in the data frame `opinion`.

Let p_j, $j = 1, 2, 3, 4, 5$ represent the true proportions of customers of each store who are expected to express displeasure about the proposed change. Then, using the contrast $\psi = 3p_1 + 3p_2 - 2p_3 - 2p_4 - 2p_5$, the hypotheses to be tested are as previously stated, and the needed calculations are taken care of as follows. First define the contrast vector

```
> k <- c(3, 3, -2, -2, -2)
```

Next, get n_j, $\hat{p}_j$, $\hat{q}_j$, $\hat{\psi}$, and $s_{\hat{\psi}}^2$ from the data.

```
> ns <- colSums(opinion)
> ps <- opinion[1, ]/ns; qs <- 1 - ps
> (tStat <- sum(k*ps)); tStatVar <- sum(k^2*ps*qs/ns)
[1] 0.6983936
```

Finally, calculate the test statistic and the p-value, remembering that $\psi_0 = 0$.

```
> chiSq <- tStat^2/tStatVar
> pchisq(q = chiSq, df = 4, lower.tail = FALSE)
[1] 0.01142841
```

Since p-value $= 0.0114$ and $\hat{\psi} = 0.6984 > 0$, one may conclude (at $\alpha \geq 0.012$) that a larger proportion of customers of Stores 1 and 2 care, and would most likely be upset, about the proposed change than customers of the other three stores.

12.3.2 Cohen's Small Sample Approach

Following the approach described in [25], with $c_1, c_2, \ldots, c_J$ being as defined for the previously described approach, using Fisher's arcsine transformation to get $\hat{\phi}_j$ in

$$\hat{\psi} = c_1 \hat{\phi}_1 + c_2 \hat{\phi}_2 + \cdots + c_J \hat{\phi}_J,$$

and defining

$$s_{\hat{\psi}}^2 = c_1^2/n_1 + c_2^2/n_2 + \cdots + c_J^2/n_J,$$

a $(1 - \alpha) \times 100\%$ confidence interval of ψ is given by

$$\hat{\psi} - \sqrt{\chi_\alpha^2}\, s_{\hat{\psi}} < \psi < \hat{\psi} + \sqrt{\chi_\alpha^2}\, s_{\hat{\psi}},$$

where χ_α^2 is as before. The hypotheses associated with the test corresponding to the above interval are as before, and the null hypothesis is rejected if ψ_0 does not lie in this interval. Also, as before, the statistic corresponding to this test is

$$\chi^{2*} = \frac{\left(\hat{\psi} - \psi_0\right)^2}{s_{\hat{\psi}}^2} \sim \chi^2(J - 1)$$

and p-value $= P(\chi^2 \geq \chi^{2*})$.

Example

Applying this approach to the data for the previous example, the following code does the job. First get the transformed proportions

```
> phi <- ifelse(test = (ps != 0) & (ps != 1),
+     yes = 2*asin(sqrt(ps)),
+       no = ifelse(test = (ps == 0),
+         yes = 2*asin(sqrt(1/(4*ns))),
+           no = pi - 2*asin(sqrt(1/(4*ns)))))
```

Then calculate $\hat{\psi}$, $s_{\hat{\psi}}$ and χ^{2*} to get p-value $= 0.0111$ and $\hat{\psi} = 1.4682 > 0$. This does not contradict the previous result from Marascuilo's method.

A Caution When Using Cohen's Approach

For simple contrasts (pairwise comparisons) it is clear that

$$p_j = p_k \quad \Leftrightarrow \quad 2\arcsin\sqrt{p_j} = 2\arcsin\sqrt{p_k} \quad \Leftrightarrow \quad \phi_j = \phi_k.$$

So $\psi = p_j - p_k$ and $\psi = \phi_j - \phi_k$ are in fact equivalent contrasts and, for the hypotheses in question, a rejection (or non-rejection) of the null hypothesis means the same thing in the transformed and untransformed setting. This is not the case if the linear contrast is not simple. However, in [25] it is stated that serious inacuracies in conclusions will not occur if the proportions lie within a small range outside the interval $(0.25, 0.75)$.

12.4 The Chi-Square Goodness-of-Fit Test

Suppose it is hypothesized that a given population is spread across J classes according to a certain distribution, say $p_1 = P_1$, $p_2 = P_2$, $\ldots$, $p_J = P_J$, where p_j denotes the true proportion of the population that falls within the jth class and P_j the hypothesized proportion. The *chi-square goodness-of-fit test* can be used to determine whether the true proportions are significantly different from hypothesized proportions. Samples for use in this test comprise a set of randomly observed frequencies, say $o_1, o_2, \ldots, o_J$, and the hypotheses to be tested have the general form

$$H_0 \quad : \quad p_1 = P_1,\ p_2 = P_2,\ \ldots,\ p_J = P_J,\quad \text{vs.}$$
$$H_1 \quad : \quad \text{The proportions do not fit the given distribution.}$$

Here, if the samples are random, the necessary computations for performing the chi-square goodness-of-fit test are as follows. Let $n = \sum_{j=1}^{J} n_j$, then each expected frequency e_j is computed using $e_j = P_j\, n$ and the test statistic is computed using

$$\chi^{2*} = \sum_{j=1}^{J} \frac{(o_j - e_j)^2}{e_j} \sim \chi^2(J - 1),$$

where $e_j \geq 5$ and p-value $= P(\chi^2 \geq \chi^{2*})$.[5] As applicable to this test, the `chisq.test` function call has the form

```
chisq.test(x, p)
```

where `x` is assigned the vector of observed frequencies o_j and `p` the vector of hypothesized proportions P_j, the default values of `p` being $P_j = 1/J$.

Example: Testing for Identical Preferences

There are four stores in the Juneau area from which residents of the Mendenhall Valley routinely get their groceries: Safeway, Super Bear, Fred Meyer, and Walmart. Susan wants to determine if Valley residents have a preference of stores; she is looking to pick a store at which she might sell jars of her specially pickled gumboots (a large mollusk commonly found in the waters of the Inside Passage, and an area delicacy).

She surveyed 250 individuals from different randomly selected homes in the Valley, the responses are given below.

Store	Safeway	Super Bear	Fred Meyer	Walmart
Observed	71	83	69	27

To test the hypothesis "Valley residents do not have store preferences," place the observed frequencies in a vector,

---

[5] The contingency table of observed frequencies for such cases has 1 row and J columns. For a computed statistic to remain constant, one observation is fixed and the rest are allowed to vary. So, the chi-square distribution associated with the statistic has $\nu = J - 1$ degrees of freedom.

```
> obsFreq <- c(71, 83, 69, 27)
```

and with the default setting for p, run

```
> chisq.test(x = obsFreq)
```

to get $\chi^{2*} = 28.72$, $\nu = 3$, and p-value = 0. Thus, there is sufficient evidence to conclude that Valley residents do have a preference when considering stores from which to get their groceries.

Example: Testing for Distinct Preferences

For the previous example, supposed it is hypothesized that of all Valley residents, 10% prefer Walmart ($P_W = 0.1$), whereas preferences are equally spread across Safeway, Super Bear and Fred Meyer ($P_S = P_{SB} = P_{FM} = 0.3$). Then, to test the hypotheses

$$H_0 \quad : \quad P_S = P_{SB} = P_{FM} = 0.3 \text{ and } P_W = 0.1 \quad \text{vs.}$$
$$H_1 \quad : \quad \text{The proportions do not fit the given distribution}$$

place the hypothesized proportions in a vector,

```
> P <- c(0.3, 0.3, 0.3, 0.1)
```

and run

```
> chisq.test(x = obsFreq, p = P)
```

to get $\chi^{2*} = 1.7067$, $\nu = 3$ and p-value = 0.6355. So, it appears Susan might be free to go with her choice of Safeway, Super Bear or Fred Meyer to market her pickled gumboots.

13

Tests of Variances and Spread

In preliminary applications of elementary statistics there are two scenarios that are typically encountered with respect to variances; the first being when the variances themselves are of interest, and the second is when the nature of the variances determines the method that may be applied to estimating and/or testing means. To this end, methods for constructing interval estimates (where appropriate) and testing hypotheses associated with population variances based on observed data for continuous random variables are covered. Code for all examples and additional explorations are contained in the file Chapter13Script.R.

13.1   Relevant Probability Distributions

Two types of probability distributions find use here, chi-square distributions and F distributions. An introduction to chi-square distributions along with relevant R functions is given in Section 11.1.4. This section provides a review of the definitions and relevant properties of the F distribution, corresponding R functions, and an outline of methods for assessing the extent and manner in which the underlying random variable for a given sample might deviate from normality.

13.1.1   F Distributions

Let X be a continuous random variable for which the sample space is the set of all positive real numbers. Let ν_1 and ν_2 be positive integers, and define the *probability density function of X* by

$$f(X) = \frac{\Gamma\left[(\nu_1 + \nu_2)/2\right]}{\Gamma\left[\nu_1/2\right]\Gamma\left[\nu_2/2\right]} \left(\frac{\nu_1}{\nu_2}\right)^{\nu_1/2} X^{\nu_1/2-1}\left(1 + \frac{\nu_1}{\nu_2}X\right)^{-(\nu_1+\nu_2)/2}$$

where $\Gamma\left[\kappa\right]$ is the gamma function mentioned in Section 11.1.4. Then X is said to have an F distribution with parameters ν_1 and ν_2,[1] and notation used

---

[1]In applications of this probability distribution the parameters ν_1 and ν_2 are referred to as the degrees of freedom of the numerator and denominator, respectively.

to indicate this will be $X \sim F(\nu_1, \nu_2)$. For $\nu_2 > 2$, the mean of X is given by $\mu = \nu_2/(\nu_2 - 2)$ and, if $\nu_2 > 4$, the variance is

$$\sigma^2 = \frac{2\nu_2^2 \left(\nu_1 + \nu_2 - 2\right)}{\nu_1 \left(\nu_2 - 2\right)^2 \left(\nu_2 - 4\right)}.$$

The R function corresponding to this density function is

```
df(x, df1, df2)
```

The argument x may be assigned a single value or a vector of values, and df1 and df2 are assigned the values ν_1 and ν_2. The remaining arguments are omitted from function calls (left in their default settings).

The *cumulative distribution function* for the F distribution is

$$P(X \leq x) = \int_0^x f(t)dt,$$

and the R function for this cumulative distribution function is

```
pf(q, df1, df2, lower.tail = FALSE)
```

Here too q may be assigned a single value or a vector of values and the purpose of the argument lower.tail is as before.

Finally, to solve equations of the form $P(X \leq x) = p$ or $P(X > x) = p$ use the function

```
qf(p, df1, df2, lower.tail = FALSE)
```

The arguments and their purposes are as previously described.

Examples

Let $X \sim F(7, 13)$, then the probability density curve is shown in Figure 13.1; see the script file for code used to construct this figure.

The function calls

```
> pf(q = 1.7, df1 = 7, df2 = 13, lower.tail = TRUE)
> pf(q = 1.7, df1 = 7, df2 = 13, lower.tail = FALSE)
```

yield $P(X \leq 1.7) = 0.8061$ and $P(X > 1.7) = 0.1939$, and an alternate option to find $P(1 \leq X \leq 2) = 0.3401322$ through the use of the diff function is

```
> diff(pf(q = c(1,2), df1 = 7, df2 = 13))
```

To find x for which $P(X \leq x) = 0.05$, run

```
> qf(p = 0.05, df1 = 7, df2 = 13, lower.tail = TRUE)
```

and x for which $P(X > x) = 0.05$, run

```
> qf(p = 0.05, df1 = 7, df2 = 13, lower.tail = FALSE)
```

As with all continuous probability distributions, individual quantiles do not contribute to cumulative probabilities. For example, $P(1 \leq X \leq 2)$ and $P(1 < X < 2)$ yield the same probability.

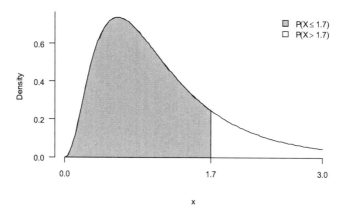

FIGURE 13.1
Density plot of $X \sim F(7, 13)$. The probability $P(X \leq 1.7)$ is given by the area of the shaded region.

13.1.2 Using a Sample to Assess Normality

Under certain conditions the random variables associated with sampling distributions of sample variances used in estimating or performing tests on population variances have chi-square or F distributions.

For the majority of procedures demonstrated in this chapter, the assumption is that the samples involved come from populations whose underlying random variables are (at least approximately) normally distributed. Therefore, it is useful to be able to assess whether samples might provide evidence that the underlying random variables deviate from normality. Two approaches to assessing normality are outlined here.

Example

See Section 8.4 for details on the construction and interpretation of normal probability QQ-plots. Here, only an illustration of their application to assessing the assumption of normality of an underlying random variable is provided.

Consider assessing the normality of the underlying random variable for the data contained in the vector **handbrake**, see Section 13.2.2 for the story behind these data. The normal probability QQ-plot along with the QQ-line, see Figure 13.2, is constructed using

```
> qqnorm(y = handbrake); qqline(y = handbrake, lty = 2)
```

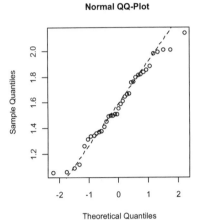

FIGURE 13.2
Normal probability QQ-plot along with the corresponding QQ-line of the sample `handbrake`.

While some gaps appear in the plotted points in Figure 13.2 and the tails are a little "off," the middle 50% of the plotted points follow the QQ-line quite closely. This plot does not suggest (at least in a strong sense) that the data in `handbrake` are associated with a non-normal random variable.

There are several tests of univariate normality that can be used (see, for example, [128]). The null hypotheses for one such test, the *Shapiro–Wilk test* [122], is that the underlying random variable is normally distributed. Computational details for this test are not provided; however, to perform the computations, run

```
> shapiro.test(x = handbrake)
```

The result of this test ($W = 0.9732$, p-value $= 0.502$) does not contradict the findings from the QQ-plot, at least for any $\alpha \leq 0.5$.

13.2   Single Population Variances

Let X represent a random variable with variance σ^2, and let s^2 be the sample variance for a sample of size n. If X is normally distributed then

$$\frac{(n-1)\,s^2}{\sigma^2} \sim \chi^2(n-1),$$

and the resulting sampling distribution finds use in estimating σ^2 and performing tests of σ^2 against hypothesized values.

For this section basic code to perform the relevant computations are presented. This code is packaged in the function `oneVar.test`, the usage definition of which is

```
oneVar.test(x, sigmaSq = 0, conf.level = 0.95,
    alternative = "two.sided")
```

where `x` is assigned the sample and `sigmaSq` the hypothesized variance. If `sigmaSq` is omitted from a function call, only a confidence interval is computed. For hypothesis tests (only) the argument `alternative` can be assigned `"less"`, `"two.sided"` (the default), or `"greater"`.

13.2.1 Estimating a Variance

If X is a normally distributed random variable and s^2 is the variance of a sample of size n, a $100(1 - \alpha)\%$ confidence interval for σ^2 is computed using

$$\frac{(n-1)\,s^2}{\chi^2_{\alpha/2}} < \sigma^2 < \frac{(n-1)\,s^2}{\chi^2_{1-\alpha/2}},$$

where the quantiles $\chi^2_{\alpha/2}$ and $\chi^2_{1-\alpha/2}$ satisfy the right-tailed probabilities $P(\chi^2 \geq \chi^2_{\alpha/2}) = \alpha/2$ and $P(\chi^2 \geq \chi^2_{1-\alpha/2}) = 1 - \alpha/2$, respectively, using $\nu = n - 1$ degrees of freedom.

Note that the interval estimate of the population standard deviation is obtained by taking the square roots of the above interval bounds.

Example

In Section 13.1.2 it was shown that the underlying random variable for the data in `handbrake` may be assumed approximately normally distributed. To obtain a 90% confidence interval of the variance of the underlying random variable for these data, run

```
> oneVar.test(x = handbrake, conf.level = 0.90)
```

to get $0.0581 < \sigma^2 < 0.1274$. Taking the square-root of the components of this inequality then gives $0.24 < \sigma < 0.36$, the 90% confidence interval of the population standard deviation.

13.2.2 Testing a Variance

Denote the hypothesized variance by σ_0^2, then to test any of the alternative hypotheses

$$H_1 : \sigma^2 < \sigma_0^2, \qquad H_1 : \sigma^2 \neq \sigma_0^2, \qquad H_1 : \sigma^2 > \sigma_0^2$$

the test statistic is

$$\chi^{2*} = \frac{(n-1)\,s^2}{\sigma_0^2}.$$

and, under the null hypothesis, $\chi^{2*} \sim \chi^2(n-1)$. For one-tailed tests the p-value is given by

$$\text{p-value} = \begin{cases} P(\chi^2 \le \chi^{2*}) & \text{for a left-tailed test} \\[2ex] P(\chi^2 \ge \chi^{2*}) & \text{for a right-tailed test} \end{cases}$$

and for a two-tailed test the p-value is computed using

$$\text{p-value} = \begin{cases} 2\,P(\chi^2 \le \chi^{2*}) & \text{if } P(\chi^2 \le \chi^{2*}) < 0.5 \\[2ex] 2\,P(\chi^2 \ge \chi^{2*}) & \text{otherwise.} \end{cases}$$

Note that this test also covers that of the population standard deviation.

Example

Here is the story behind the **handbrake** data. Two friends, "Handbrake" Harry and "Leadfoot" Larry, got into a heated discussion about Handbrake's reflexes.[2] Leadfoot insisted that Handbrake's reflexes are all over the place; that is, his reflex times have a high standard deviation. Handbrake, on the other hand, claims that the true standard deviation of his reflexes is less than 0.5 seconds. They decide to test Handbrake's claim by gathering some data;[3] these are contained in the numeric vector **handbrake**.

Let σ^2 represent the true variance of Handbrake's reflex times. Then, the equivalent hypotheses to be tested are

$$H_0 : \sigma^2 \ge 0.25 \quad \text{vs.} \quad H_1 : \sigma^2 < 0.25 \quad \text{(Claim)},$$

and the code

```
> oneVar.test(x = handbrake, sigmaSq = 0.5^2,
        alternative = "less")
```

produces $\chi^{2*} = 11.854$, $df = 36$, p-value $= 0.00005$. So, for any $\alpha > 0.00005$, there is sufficient evidence to support Handbrake's claim that the true standard deviation of his reflexes is less than 0.5 seconds.

---

[2] These nicknames are well-earned; however, the story behind them is a long one and is not relevant to the problem in question.

[3] They were unwilling to reveal anything about the experiment, other than the fact that a statistician friend designed it, and that it was conducted in the vicinity of the MacNugget intersection (in Juneau, AK).

13.3   Exactly Two Population Variances

Let X_1 and X_2 represent two random variables with unknown variances σ_1^2 and σ_2^2. Let s_1^2 and s_2^2 be the sample variances for two samples of size n_1 and n_2 for these random variables, respectively. If X_1 and X_2 are independent, and (at least approximately) normally distributed, then

$$\frac{s_1^2 / \sigma_1^2}{s_2^2 / \sigma_2^2} \sim F(n_1 - 1, n_2 - 1),$$

and this sampling distribution finds use in estimating or performing tests on the ratio of the variances σ_1^2 and σ_2^2.

The R function, contained in package `stats`, that finds use for all computations in this section is `var.test`. The usage definition of this function has two formats. The first is designed to take in the data in two separate vectors and has the form

```
var.test(x, y, data, alternative = "two.sided",
    conf.level = 0.95)
```

where `x` represents the first sample and `y` the second. The second application finds use when the data are stored in bivariate format, one variable representing the numeric data and the other a factor that contains as its levels the sample identifiers. For this form the data are entered as a formula.

```
var.test(formula, data, alternative = "two.sided",
    conf.level = 0.95)
```

The argument `formula` is assigned an expression having the general form `numericVar ~ groupingVar`, the two variables being contained in the data frame assigned to the argument `data`. The arguments `alternative` and `conf.level` are as from previous test functions.

13.3.1   Estimating the Ratio of Two Variances

The quantity assessed here is a ratio, specifically, the ratio σ_1^2/σ_2^2. If X_1 and X_2 are independent and normally distributed random variables and s_1^2 and s_2^2 are sample variances for two samples of size n_1 and n_2 for these random variables, then a $100(1-\alpha)\%$ confidence interval for σ_1^2/σ_2^2 is computed using

$$\frac{1}{F_{\alpha/2}} \frac{s_1^2}{s_2^2} < \frac{\sigma_1^2}{\sigma_2^2} < \frac{1}{F_{1-\alpha/2}} \frac{s_1^2}{s_2^2},$$

where $P(F \geq F_{1-\alpha/2}) = 1 - \alpha/2$ and $P(F \geq \alpha/2) = \alpha/2$ with $\nu_1 = n_1 - 1$ and $\nu_2 = n_2 - 1$.[4]

Example

A popular fish for some salt-water anglers in the Juneau area is the Dolly Varden (affectionately called Dolly). People who fish for Dollies have their favorite spots and, very typically, do not switch spots very frequently. "Doc" Dalton Dobson is a newcomer to the Juneau area and is also an avid Dolly fisherman. He has decided to approach the spot-selection process scientifically. He is a shore-fisherman and reasons that there are two kinds of salt-water fishing locations: those near the mouths of streams, and those that are not.

By eavesdropping on conversations at some popular coffee shops and striking up conversations with confessed Dolly-enthusiasts, he compiled a list of all the area fishing spots that can be reached from the shore and that fit into one or the other of his classifications (next to a stream, or not next to a stream). From this list he randomly selected one convenient location from each type of spot: by the Douglas Bridge (which is next to some streams), and Point Louisa (which is not next to any streams).

As soon as Dollies started appearing in salt-water he went to the Douglas Bridge location two hours before high tide every day for a week, and fished until high tide. The following week he went to the Point Louisa location and did the same. Since there is a catch limit on Dollies he practiced catch-and-release, weighing each and every fish (in lbs) before releasing it. He also recorded the rod-time in minutes for each fish caught using a special formula he obtained from *The Statisticians Handbook to Fishing*.

The data for his study are contained in the data frame `dolly`, and are recorded under the variables `weight`, `time` and `where`. The levels of `where` are `DB` (Douglas Bridge, level 1) and `PL` (Point Louisa, level 2).

Doc's question, relevant to this section, addresses estimating the ratio of the variances of fish sizes from the two locations. The two samples, being collected from distinct locations that are quite far apart, are considered independent. Moreover, a preliminary analysis of the data suggests the assumption of approximate normality on the underlying random variables is not violated; see the script for this example.

Let σ_{DB}^2 denote the variance of the weights of Dollies found at the Douglas Bridge location and σ_{PL}^2 the variance of the weights of Dollies found at the Point Louisa location, and consider obtaining a 90% confidence interval of $\sigma_{DB}^2/\sigma_{PL}^2$. Among the output produced by the function `var.test` is the

---

[4]Note that in texts, for ease in using tables, this is more typically written in the form

$$\frac{s_2^2}{s_1^2} \, F_{1-\alpha/2} < \frac{\sigma_2^2}{\sigma_1^2} < \frac{s_2^2}{s_1^2} \, F_{\alpha/2},$$

where s_2^2 is chosen to be the larger variance; see for example [7, p. 358]. However, the function `var.test` produces the bounds for the reciprocal of this ratio and the relative sizes of the two sample variances are not of concern.

confidence interval (as `conf.int`) corresponding to the `alternative` entered into the function. Thus, running

```
> var.test(formula = weight ~ where,
+      data = dolly, conf.level = 0.90)$conf.int
```

outputs the desired two-sided 90% confidence interval which, rounded to 4 decimal places, is $0.5388 < \sigma_{\mathrm{DB}}^2/\sigma_{\mathrm{PL}}^2 < 1.5592$.

13.3.2 Testing the Ratio of Two Variances

The alternative hypotheses relevant to this section are[5]

$$H_1 : \sigma_1^2/\sigma_2^2 < 1, \qquad H_1 : \sigma_1^2/\sigma_2^2 \neq 1, \qquad H_1 : \sigma_1^2/\sigma_2^2 > 1.$$

If X_1 and X_2 are independent and normally distributed random variables and s_1^2 and s_2^2 are sample variances for two samples of size n_1 and n_2 for these random variables, then under the null hypothesis

$$F^* = s_1^2/s_2^2$$

has an F-distribution with $\nu_1 = n_1 - 1$ and $\nu_2 = n_2 - 1$ degrees of freedom.[6]

The p-value corresponding to the alternative is then computed as follows. For one-tailed tests use

$$\text{p-value} = \begin{cases} P(F \leq F^*) & \text{for a left-tailed test} \\ P(F \geq F^*) & \text{for a right-tailed test} \end{cases}$$

and use

$$\text{p-value} = \begin{cases} 2\,P(F \leq F^*) & \text{if } P(F \leq F^*) < 0.5 \\ 2\,P(F \geq F^*) & \text{otherwise.} \end{cases}$$

for a two-tailed test.

Example

Continuing with Doc's Dolly study, consider testing the hypotheses

$$H_0 : \sigma_{\mathrm{DB}}^2/\sigma_{\mathrm{PL}}^2 = 1 \quad \text{vs.} \quad H_1 : \sigma_{\mathrm{DB}}^2/\sigma_{\mathrm{PL}}^2 \neq 1.$$

at the $\alpha = 0.10$ level of significance. To do this, run

---

[5]Note that $\sigma_1^2/\sigma_2^2 = 1$ and $\sigma_1^2 = \sigma_2^2$ are equivalent statements, so these alternative hypotheses are equivalent to

$$H_1 : \sigma_1^2 < \sigma_2^2, \qquad H_1 : \sigma_1^2 \neq \sigma_2^2, \qquad H_1 : \sigma_1^2 > \sigma_2^2.$$

[6]For convenience in using tables to find p-values, some elementary statistics texts recommend that the test statistic always be computed by placing the sample with the larger sample variance in the numerator; that is, name the variable with the larger sample variance X_1. *This strategy is not used here* since R takes care of computing p-values.

```
> var.test(formula = weight ~ where,
+       data = dolly, conf.level = 0.90)
```

to get $F^* = 0.8997$, $df_N = 32$, $df_D = 51$ and p-value $= 0.7614$. So, at any (typical) level of significance there is insufficient evidence to conclude there is a difference between the true variances of Dolly weights from the two locations.

A Comment about the Confidence Interval Produced: The 90% confidence interval produced, $0.5388 < \sigma^2_{DB}/\sigma^2_{PL} < 1.5592$, contains 1, so this suggests (at $\alpha = 0.10$) that there is insufficient evidence to conclude that the ratio of the two variances differs from 1.

13.3.3 What If the Normality Assumption Is Violated?

Alternative two-sample tests for dispersion exist, and R functions for two such methods are available in package stats. The two methods are the *Ansari–Bradley two sample test* (ansari.test) and *Mood's two-sample test* (mood.test), see the script file for code applied to the Dolly data. The next section includes three *Levene-type procedures* and one rank-based procedure, all of which are robust to deviations from normality and that can be used to test the equality of two variances. See, for example, [18], [28] and [59] for detailed discussions on these methods.

13.4 Two or More Population Variances

Let $X_1, X_2, \ldots, X_k$ be $k \geq 2$ independent random variables with unknown variances $\sigma^2_1, \sigma^2_2, \ldots, \sigma^2_k$. The hypotheses of interest in this section are

$$H_0 : \sigma^2_1 = \sigma^2_2 = \cdots = \sigma^2_k, \quad \text{vs.} \quad H_1 : \text{The variances are not all equal}$$

Four procedures for testing these hypotheses are given here. The first three methods, referred to as Levene-type procedures in [18] and discussed further in [59], have been shown to be robust to certain deviations from the normality assumption. The last method, the Fligner–Killeen test [54], is a rank-based test that can be used in cases where serious deviations from the assumption of normality are encountered.

13.4.1 Assessing Spread Graphically

A simple and reasonably informative figure for this purpose would be a stripchart, and further insight might be obtained by superimposing stripcharts on boxplots or error bar plots.[7]

---

[7]See Chapters 8 and 10 for a background review of stripcharts, boxplots and error bars.

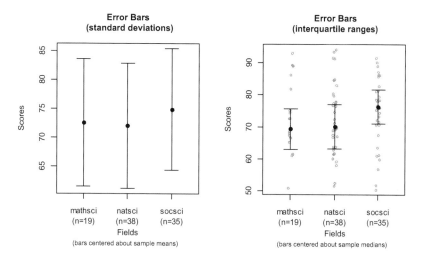

FIGURE 13.3
Using error bars and/or a stripchart to assess variability of traditional course scores by fields in the `elemStatStudy` data.

Example

Consider using the `elemStatStudy` data from previous illustrations. To perform a visual exploration of the spread of traditional course scores by fields the function `errorBar.plot` is used to construct the plots in Figure 13.3; see Section 10.5 and code in the script file.

Both the plot of standard deviation error bars, and the plot of interquartile range error bars suggest fairly equivalent spreads of observed scores for all three samples. The stripchart included in the right-hand plot provides a feel for how individual scores are clustered.

13.4.2 Levene's Test

Originally proposed in [81] and later extended in [42, pp. 38–39], *Levene's test* can be used if the underlying random variables have symmetric distributions that deviate from normality.

For $j = 1, 2, \ldots, k$ let n_j be the size of sample j, and for $i = 1, 2, \ldots, n_j$ let x_{ij} be the ith observation for the jth sample. Denote the mean of the jth sample by $\bar{x}_j$ and compute the absolute deviations

$$d_{ij} = |x_{ij} - \bar{x}_j|.$$

Let $n = \sum_{j=1}^{k} n_j$, denote the mean of all the absolute deviations by $\bar{d}$, and the sample means and variances of the absolute deviations by $\bar{d}_j$ and $s_{d_j}^2$, respectively. Then the test statistic is

$$F^* = \frac{\sum_{j=1}^{k} n_j \left(\bar{d}_j - \bar{d} \right)^2 \big/ (k-1)}{\sum_{j=1}^{k} (n_j - 1) s_{d_j}^2 \big/ (n-k)},$$

where $F^* \sim F(k-1, n-k)$, and p-value $= P(F \geq F^*)$.[8]

The computations for this, and the next two Levene-type tests (see, also, Section 6.5) are contained in the function `leveneType.test`. For the test under discussion here, the usage definition for this function is

```
leveneType.test(x, g, center = "mean")
```

where x and g are as defined previously, and `center = "mean"` instructs the function to use deviations from the sample means in its computations.

Example

To apply this test to the same data as for the previous example using the `leveneType.test` function, run

```
> with(data = subset(x = elemStatStudy,
+            subset = (course == "trad")),
+      expr = leveneType.test(x = score, g = field,
+            center = "mean"))
```

to get $F^* = 0.0808$, $df_1 = 2$, $df_2 = 89$, p-value $= 0.9225$ which indicates that there is insufficient evidence to reject an assumption of equal variances of scores in the traditional course across fields.

13.4.3 Levene's Test with Trimmed Means

Deviations from normality in the underlying random variables may occur by way of symmetric long-tailed distributions. In such cases, the modification of Levene's test suggested in [18] is to use the 10% trimmed mean in place of the sample mean $\bar{x}_j$.

Thus, if $\check{x}_j$ denotes the 10% trimmed mean for the jth sample, the only modification to Levene's statistic is in how the absolute deviations d_{ij} are calculated, that is, for this procedure

$$d_{ij} = \left| x_{ij} - \check{x}_j \right|.$$

The remaining calculations are the same as for Levene's test, and the code

---

[8]This test statistic is the result of performing a one-way analysis of variance (ANOVA) of these absolute deviations across the k samples; see Section 15.3.

```
leveneType.test(x, g, center = "trimmed.mean")
```

instructs the function `leveneType.test` to perform this procedure.

Example

For the same data, run the `leveneType.test` function with the argument assignment `center = "trimmed.mean"` to get $F^* = 0.1087$, $df_1 = 2$, $df_2 = 89$, p-value = 0.8971. The result agrees with those of the previous method.

13.4.4 Brown–Forsythe Test

To combat assymmetry in the distributions of the underlying random variables, Brown and Forsythe suggested using the sample median in place of the mean in Levene's test, see [18]. This modification has come to be known as the *Brown–Forsythe test*.

So, denoting the median of the jth sample by $\tilde{x}_j$, the alteration to Levene's test is to first compute the absolute deviations d_{ij} using

$$d_{ij} = |x_{ij} - \tilde{x}_j|,$$

and then duplicate the calculations for the previous two procedures. The code

```
leveneType.test(x, g, center = "median")
```

instructs the function `leveneType.test` to perform the Brown–Forsythe procedure. This is the default procedure for `leveneType.test`.

Example

Using the same data again, run the `leveneType.test` function with `center = "median"` to get $F^* = 0.0726$, $df_1 = 2$, $df_2 = 89$, p-value = 0.9300. Yet again, there is insufficient evidence to suggest that the true variances of the scores in the traditional course differ across fields.

13.4.5 Fligner–Killeen Test

This method, a modification/extension of the procedure proposed by Fligner and Killeen in [54] (see [28] for details), makes use of the ranks of absolute deviations from sample medians.

Denote the median of the jth sample by $\tilde{x}_j$ and, as for the previously described Brown–Forsythe method, compute

$$d_{ij} = |x_{ij} - \tilde{x}_j|.$$

Next, obtain the ranks, R_{ij}, of the combined absolute deviations d_{ij} from smallest to largest (using mean ranks for ties) and compute standard normal scores (quantiles) corresponding to these ranks by finding a_{ij}, that satisfy

$$P\left(Z \leq a_{ij}\right) = \frac{1 + R_{ij}/(n+1)}{2}.$$

Now denote the mean score for the jth sample by $\bar{a}_j$ and the overall mean score by $\bar{a}$, then the test statistic is computed using

$$\chi^{2*} = \frac{\sum_{j=1}^{k} n_j \left(\bar{a}_j - \bar{a}\right)^2}{s_a^2}$$

where s_a^2, the variance of the a_{ij} is given by

$$s_a^2 = \sum_{j=1}^{k} \sum_{i=1}^{n_j} \left(a_{ij} - \bar{a}\right)^2 \bigg/ (n-1).$$

Finally, with $\chi^{2*} \sim \chi^2(k-1)$, p-value $= P(\chi^2 \geq \chi^{2*})$ is computed.

The R function for this procedure is `fligner.test`, and its usage definition is either of

```
fligner.test(x, g, ...)
fligner.test(formula, data, subset, ...)
```

where x is assigned the vector of observed values and g the corresponding factor containing the groupings of the entries of x. For the formula method, an argument assignment of the form `formula = x ˜ g` is used.

Example

Applying this method to the previous example,

```
> fligner.test(formula = score ˜ field,
+      data = elemStatStudy, subset = (course == "trad"))
```

gives $\chi^{2*} = 0.0947$, $df = 2$, p-value $= 0.9538$. Here too there is insufficient evidence to suggest that the true variances of the scores in the traditional course differ across fields.

14

Tests for One or Two Means

This chapter addresses commonly encountered methods of estimating and testing one population mean or the difference between two population means that are associated with random variables that are (at least approximately) normally distributed. For larger samples, as a consequence of the Central Limit Theorem, the methods described here apply irrespective of whether the underlying random variables are normally distributed or not. Additionally, the previous chapter's methods are relevant since, in comparing the means of two independent populations, the computations involved are determined by whether or not the population variances are assumed equal. Code for illustrations in this chapter are contained in the script file `Chapter14Script.R`.

14.1 Student's t Distribution

The probability distribution already seen that finds use in this chapter is the standard normal distribution; see Section 11.1.3. The primary distribution needed here is *Student's t distribution*.

Let X be a continuous random variable for which the sample space is the set of all real numbers. Let ν be a positive integer and define the probability density function of X by

$$f(X) = \frac{1}{\sqrt{\nu\pi}} \frac{\Gamma\left[(\nu+1)/2\right]}{\Gamma\left[\nu/2\right]} \left(1 + \frac{X^2}{\nu}\right)^{-(\nu+1)/2}.$$

Then X is said to be *t distributed with the parameter* ν.[1] Notation used to indicate this will be $X \sim t(\nu)$. In the case where $\nu > 1$, X has mean $\mu = 0$ and for $\nu > 2$ the variance of X is $\sigma^2 = \nu/(\nu-2)$.

The R function corresponding to this probability density function is

```
dt(x, df)
```

The arguments `x` and `df` have the same meaning and purposes, and are used

---

[1] As for the chi-squared distribution ν is called the degrees of freedom of the t-distribution in question, and Γ represents the gamma function described in Footnote 3 of Section 11.1.4.

in the same manner as for R functions for the chi-square distribution. The cumulative distribution function is defined by

$$P(X \leq x) = \int_{-\infty}^{x} f(t)\, dt,$$

and, being a continuous probability distribution,

$$P(X \leq x) = P(X < x) \quad \text{and} \quad P(X \geq x) = P(X > x).$$

The R function for computing this integral for a given $X = x$ is

```
pt(q, df, lower.tail = TRUE)
```

The arguments q, df and lower.tail serve the same purposes as already described for earlier seen such functions. The function,

```
qt(p, df, lower.tail = TRUE)
```

is the one to use when an approximate quantile corresponding to a given left- or right-tailed probability, p, is needed.

Examples

The graph of the density function for $X \sim t(11)$ in Figure 14.1 was constructed using the dt function (see script file for code).[2]

To calculate the left-tailed probability $P(X \leq 1.35) = 0.89793$ or the right tailed probability $P(X > 1.35) = 0.10207$, run

```
> pt(q = 1.35, df = 11, lower.tail = TRUE)
> pt(q = 1.35, df = 11, lower.tail = FALSE)
```

and to calculate the probability $P(-1 \leq X \leq 1)$, run

```
> diff(pt(q = c(-1,1), df = 11))
```

To compute the quantile x for which $P(X \leq x) = 0.01$, run

```
> qt(p = 0.01, df = 11, lower.tail = TRUE)
```

and run

```
> qt(p = 0.01, df = 11, lower.tail = FALSE)
```

to find x for which $P(X > x) = 0.01$.

---

[2] Like the standard normal distribution the t-distribution has a bell-shaped density curve that is symmetric about 0. This distribution differs from the standard normal distribution in that it has a variance $\sigma^2 = \nu/(\nu - 2) > 1$. However, since

$$\lim_{\nu \to \infty} \left(\frac{\nu}{\nu - 2} \right) = 1,$$

as ν becomes large the corresponding t-distribution approaches the standard normal distribution. In such cases the standard normal distribution can be used as an approximation of the t-distribution.

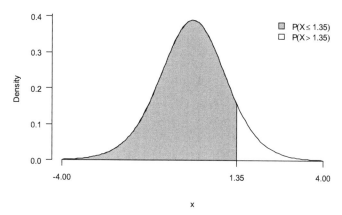

FIGURE 14.1
Density curve of $X \sim t(11)$. The shaded area represents a left-tailed probability and the unshaded area a right-tailed probability.

14.2 Single Population Means

Let X represent a random variable with mean μ and standard deviation σ and let $\bar{X}$ represent the sample mean for any sample of size n from the population for which X is the underlying random variable.

Now assume one of two possible scenarios: Either n is large, in which case the sampling distribution of $\bar{X}$ is approximately normally distributed; or X is normally distributed, in which case the sampling distribution of $\bar{X}$ is also normally distributed.

In both cases $\mu_{\bar{X}} = \mu$ and $\sigma_{\bar{X}} = \sigma/\sqrt{n}$. So, under these two scenarios $\bar{X} \sim N(\mu_{\bar{X}}, \sigma_{\bar{X}})$ is at least approximately the case and for a given sample standard deviation s,

$$\frac{\bar{X} - \mu}{s/\sqrt{n}} \sim t(n-1).$$

This sampling distribution is relevant to applications for this section.

The usage definition and relevant arguments for the R function that applies to the majority of this section, with arguments of interest, is

```
t.test(x, data, alternative, mu, conf.level = 0.95)
```

Here x is assigned the sample in question, which may be contained in the data frame or list assigned to data. The argument data may be excluded from the function call if the contents of the vector assigned to x can be accessed directly. The arguments alternative and conf.level are as previously defined and, when relevant, mu is assigned the hypothesized mean, the default value of which is zero.

14.2.1 Verifying the Normality Assumption

For large samples,[3] if the underlying random variable X has mean μ and standard deviation σ, the Central Limit Theorem states that the sampling distribution of $\bar{X}$ with samples of size n is approximately normally distributed with mean $\mu_{\bar{X}} = \mu$ and standard deviation $\sigma_{\bar{X}} = \sigma/\sqrt{n}$. See code at the end of the script file for this chapter for simulations that demonstrate this. So, in such cases there is no need to verify this assumption.

For small samples ($n < 30$), however, the validity of the assumption of normality on X becomes necessary to ensure that $\bar{X}$ is (at least approximately) normally distributed. The process and code for assessing the (approximate) validity of the assumption that the underlying random variable for a given sample is normally distributed is the same as illustrated in Section 13.1.2. That is, one may use a normal probability QQ-plot for a graphical assessment or one may use a formal test such as the Shapiro–Wilk test.

14.2.2 Estimating a Mean

A $100(1 - \alpha)\%$ confidence interval for μ is computed using

$$\bar{x} - t_{\alpha/2}\ s/\sqrt{n} < \mu < \bar{x} + t_{\alpha/2}\ s/\sqrt{n}$$

where $t_{\alpha/2}$ satisfies $P(t \geq t_{\alpha/2}) = \alpha/2$ with $\nu = n - 1$ degrees of freedom.

The function t.test serves the purpose here and a typical function call, with relevant arguments, has the form

```
t.test(x, alternative = "two.sided", conf.level = 0.95)
```

The sample in question is assigned to the argument x and the conf.level argument setting can be left as is (omitted) or assigned a different confidence level. All other arguments, not shown here, are left in their default setting and, as such, are omitted from the function call.

Included in the output produced by running the function t.test is the confidence interval, identified as conf.int, that corresponds to the indicated alternative. To obtain traditional two-sided confidence intervals the default setting alternative = "two.sided" is used. So, a function call such as

```
t.test(x, conf.level = 0.95)$conf.int
```

limits the output to only the confidence interval. The argument conf.level need only be included if a confidence level different from 95% is desired.

---

[3] The commonly followed rule of thumb being to consider a sample large enough if $n \geq 30$.

Example

From the earlier Handbrake Harry example (see Section 13.2.2), consider obtaining a 90% confidence interval estimate of Handbrake's true mean reflex time. Since the sample size is greater than 30 ($n = 37$), an assessment of the normality assumption is skipped.[4] Then, running

```
> t.test(x = handbrake, conf.level = 0.90)$conf.int
```

reveals that Handbrake can be 90% confident that the interval $(1.50, 1.66)$ contains the true mean of his reflex times.

14.2.3   Testing a Mean

Denote the hypothesized mean by μ_0, then relevant alternative hypotheses for testing the true mean against this hypothesized mean include

$$H_1 : \mu < \mu_0, \qquad H_1 : \mu \neq \mu_0, \qquad H_1 : \mu > \mu_0.$$

If either of the conditions stated previously are satisfied, under the null hypothesis

$$t^* = \frac{\bar{x} - \mu_0}{s/\sqrt{n}} \sim t(n - 1),$$

and

$$\text{p-value} = \begin{cases} P(t \geq |t^*|) & \text{for one-tailed tests} \\ 2\,P(t \geq |t^*|) & \text{for two-tailed tests} \end{cases}$$

The function t.test serves the purpose again and, in this case, the arguments mu and alternative play a role.

```
t.test(x, mu, alternative = "two.sided", conf.level = 0.95)
```

The argument mu is assigned the hypothesized value and alternative is assigned the appropriate test type. Note that the confidence level plays no role in the computations of the test statistic and p-value.

Example

Now suppose that Handbrake claimed his true mean reflex time is at most 1 second. The present method can be employed to test his claim, the hypotheses being

$$H_0 : \mu \leq 1 \quad \text{(Claim)} \quad \text{vs.} \quad H_1 : \mu > 1.$$

To do this, run

---

[4]Additionally, while not necessary in this case, it was shown in Section 13.1.2 that the assumption of normality may be considered approximately satisfied.

```
> t.test(x = handbrake, mu = 1, alternative = "greater")
```

to find that $t^* = 12.323$, $df = 36$, p-value $= 0$. It appears Handbrake is way off with his claim; there is ample evidence to reject his claim at any level of significance.

A Comment about the Confidence Interval Produced: Since the argument `conf.level` was left in its default setting, the 95% confidence interval produced by `t.test` for this right-tailed test is, to two decimal places, approximately

$$1.50 < \mu < \infty.$$

The upper limit ∞ indicates the interval is associated with a right-tailed test. Using this interval, the result of the hypothesis test can be equivalently stated as follows. Since the hypothesized value, $\mu_0 = 1$, does not lie inside the confidence interval, the null hypothesis is rejected at the $\alpha = 0.05$ level of significance.

14.2.4   Can a Normal Approximation Be Used Here?

Sure, but under certain conditions. It is known that if $X \sim N(\mu, \sigma)$ and if the population standard deviation is known, then

$$\frac{\bar{X} - \mu}{\sigma/\sqrt{n}} \sim N(0, 1)$$

irrespective of the sample size.

It is not usually the case that the population standard deviation is known. However, *if* the population standard deviation *is* known (and used) then the standard normal distribution *has* to be used. To perform the previously described tasks, replace all references to the distribution $t(n-1)$ by the distribution $N(0, 1)$. Thus, the $100(1 - \alpha)\%$ confidence interval for μ is computed using

$$\bar{x} - z_{\alpha/2}\ \sigma/\sqrt{n} < \mu < \bar{x} + z_{\alpha/2}\ \sigma/\sqrt{n}$$

where $z_{\alpha/2}$ satisfies $P(Z \geq z_{\alpha/2}) = \alpha/2$ with $Z \sim N(0, 1)$. The corresponding test statistic is

$$z^* = \frac{\bar{x} - \mu_0}{\sigma/\sqrt{n}} \sim N(0, 1),$$

and the p-value is computed analogously to the previously described t-test, but using the distribution $N(0, 1)$ instead of $t(n - 1)$.

Alternatively, if the population variance is unknown but it is known that $X \sim N(\mu, \sigma)$ and the sample size is large, then

$$\frac{\bar{X} - \mu}{s/\sqrt{n}} \sim N(0, 1)$$

in the approximate sense. This approximation improves for larger and larger samples sizes; see the script for this section for a simple graphical demonstration of this effect. In this case replace σ with s in the previous z-interval and z-test statistic formulas.

A specific built-in function (of the form `z.test`) does not exist in the base R packages. However, it is not too hard to write code that will implement the above described computations; see the script file for code that repeats the previous two examples using the normal approximation approach. The numbers produced are very close to those produced by the function `t.test`.

14.3   Exactly Two Population Means

The methods described in this section concern numeric samples associated with two random variables, say X_1 and X_2, that may or may not be independent.

For small samples, it is assumed that $X_1 \sim N(\mu_1, \sigma_1)$ and $X_2 \sim N(\mu_2, \sigma_2)$. Here, an added matter of concern for independent samples is whether or not the unknown population variances σ_1^2 and σ_2^2 are assumed equal. Whatever the case, the alternative hypotheses can have one of the forms

$$H_1 : \mu_1 < \mu_2, \qquad H_1 : \mu_1 \neq \mu_2, \qquad H_1 : \mu_1 > \mu_2,$$

and the computations used vary depending on whether the underlying random variables are independent or dependent, and whether the population variances are assumed equal or not.

If both sample sizes are large (traditionally if $n_1 \geq 30$ and $n_2 \geq 30$) the normality assumption requirement on X_1 and X_2 is relaxed.

14.3.1   Verifying Assumptions

Given two random samples, the independence (or dependence) of the observations and, hence, the two underlying random variables is typically determined (and guaranteed) by the experiment design. So, for the present illustrations this will not be tested.

For the small sample scenario, the assumptions of normality on the underlying random variables X_1 and X_2 can be verified using normal probability QQ-plots or the Shapiro–Wilk test. Additionally, methods from Chapter 13 can be employed to determine whether the variances may be assumed equal, the specific method used being dependent on whether or not the normality assumption is approximately valid. Illustrations for these are contained in the code in this chapter's script file for each of the examples that follow.

14.3.2   The Test for Dependent Samples

Let X_1 and X_2 represent two *dependent* random variables; the values for each random variable are obtained from the same subject, but under different experimental conditions.[5] Since X_1 and X_2 are random variables, the paired-difference $D = X_1 - X_2$ is also a random variable having a true (unknown) mean of $\mu_D = \mu_1 - \mu_2$. The alternative hypotheses equivalent to those listed in the introduction to this section are

$$H_1 : \mu_D < 0, \qquad H_1 : \mu_D \neq 0, \qquad H_1 : \mu_D > 0.$$

Observe that this boils down to a single sample test of paired-differences with the hypothesized mean (difference) being zero.

Let $n = n_1 = n_2$ denote the common sample size (this is the case since the same subjects yield both samples) and, for $i = 1, 2, \ldots, n$, denote the ith observed paired-difference by

$$d_i = x_{i2} - x_{i1},$$

where x_{i1} and x_{i2} are the ith observed values for the two samples. Let $\bar{d}$ be the mean, and s_d the standard deviation of the resulting sample of paired-differences. If the underlying random variables are approximately normally distributed, or the paired sample is sufficiently large in size $(n \geq 30)$ then, under the null hypothesis,

$$t^* = \frac{\bar{d}}{s_d/\sqrt{n}} \sim t(n-1)$$

and

$$\text{p-value} = \begin{cases} P(t \geq |t^*|) & \text{for one-tailed tests} \\[2ex] 2\,P(t \geq |t^*|) & \text{for two-tailed tests} \end{cases}$$

using the t-distribution with $\nu = n - 1$ degrees of freedom.

Three variations of the the the `t.test` function can be used. One approach is to enter the data as a single vector of paired differences; in this case the function call has the appearance

```
t.test(x, data, alternative, conf.level)
```

where `x` is assigned the vector of paired-differences and the hypothesized value is left in its default setting of `mu = 0` (and hence omitted from the function call).

On the other hand, if the data are stored in, and passed into the function as two separate vectors of equal length, then the function call has the appearance

---

[5] For example, X_1 might represent a randomly selected student's score on a poetry writing test before listening to country music, and X_2 the same student's score on an equivalent test right after listening to an hour of country music.

```
t.test(x, y, data, alternative, conf.level, paired = TRUE)
```

where x and y are assigned the two vectors and **paired = TRUE** is required.

For the last variation, suppose the data are stored in a data frame in bivariate form: a numeric vector containing the combined samples and a factor containing the sample identifiers. Here the function call takes on the appearance

```
t.test(formula, data, alternative, conf.level, paired = TRUE)
```

where **paired = TRUE** is once again required and **formula** is used as described in its previous occurrences.

Note that the inclusion of the arguments **data**, **alternative** and **conf.level** depend on how the function is used and what it is to be used for. Which approach to use is data-format (and sometimes user) dependent.

Here is an example of the last approach to using the **t.test** function. The script file shows code for the other two approaches, both of which duplicate the results shown.

Example

"Happy" Hannah Hanson is a mental health counselor who recently attended a workshop on a therapeutic tool referred to as *Emotional Freedom Techniques* (*EFT*). The claim is that EFT helps heal (thus, lower) emotional stress, and Happy wanted to test this claim before using these techniques on her clients.

Counselors and psychologists in her city have access to a ready pool of student subjects from a fairly large nearby university; the university has a work-study arrangement for students who participate in psychological studies conducted in the surrounding area. She decides that an ideal population on which to perform a preliminary test of the techniques are students who are enrolled in the university's GER mathematics course, *and* who suffer from math and exam anxiety. At the start of the Fall 2012 semester, and with the help of the university's Student Services Office, she obtained a random sample of 73 students who were enrolled in the GER mathematics course, and who met all of the various criteria for the study.

She obtained two sets of data from the same students in the study immediately after two consecutive midterm exams For both samples she used a special instrument to measure the students' overall anxiety level during the exam; a value of 0 indicates complete calm, and a value of 10 indicates the occurence of a severe panic attack. She also recorded the students' exam scores, on a scale of 0 to 100%, for both exams. The data are contained in the data frame **eft**. Since the samples for this study are dependent, the appropriate test to compare the means (of anxiety levels or exam scores) is the t-test of means using two dependent samples (paired-samples t-test).

A quick look at the contents of the data frame **eft** shows that the levels of when are coded as "after" = 1 and "before" = 2. Knowing this is particularly important if the formula version of the function is used.[6] R sets up relevant computations according to the ordering of the levels of the factor (the level-2 anxiety values are subtracted from the level-1 anxiety values).

Since the goal is to determine whether the treatment lowers anxiety, the hypotheses to be tested can be written as (keeping in mind the order of the levels)[7]

$$H_0 : \mu_{\text{after}} \geq \mu_{\text{before}} \quad \text{vs.} \quad H_1 : \mu_{\text{after}} < \mu_{\text{before}}.$$

Since the sample size is large ($n = 73$) the normality assumption is not of concern here.

Using the formula approach, and keeping the labeling of the levels in mind, the t.test function call has the appearance

```
> t.test(formula = anxiety ~ when, data = eft,
+       alternative = "less", paired = TRUE)
```

The resulting output is $t^* = -3.7685$, $df = 72$ and p-value = 0.0001667. So, at any $\alpha > 0.0002$, there is sufficient evidence to conclude that EFT does indeed reduce anxiety, at least for the population in question.

A Comment about the Confidence Interval Produced: Since the argument conf.level was left in its default setting, the 95% confidence interval produced by t.test for this left-tailed test is, to two decimal places, approximately

$$-\infty < \mu_D < -0.49.$$

The lower limit $-\infty$ indicates a left-tailed test. Using this interval, the result of the hypothesis test can be equivalently stated as follows. Since zero, the hypothesized difference, does not lie inside the confidence interval, the null hypothesis is rejected at the $\alpha = 0.05$ level of significance.

14.3.3   Tests for Independent Samples

Let X_1 and X_2 represent two *independent* random variables with unknown population means μ_1 and μ_2. The alternative hypotheses to be tested may be one of the following.

$$H_1 : \mu_1 < \mu_2, \qquad H_1 : \mu_1 \neq \mu_2, \qquad H_1 : \mu_1 > \mu_2.$$

---

[6] Another way to be really sure of how a factor in a data frame is coded is to run something like edit(eft$when). For this particular case .Label = c("after", "before") indicates that the first level is labeled "after" and the second "before".

[7] Note that in terms of paired-differences, using $D = X_1 - X_2$ where X_1 represents anxiety levels after EFT, and X_2 represents anxiety levels before EFT, the equivalent hypotheses are

$$H_0 : \mu_D \geq 0 \quad \text{vs.} \quad H_1 : \mu_D < 0.$$

Suppose further that X_1 and X_2 are (at least approximately) normally distributed, or $n_1 \geq 30$ and $n_2 \geq 30$. There are two scenarios: when the population variances are equal, and when the population variances are unequal. Unless there is sound justification for assuming the population variances are equal, they should be assumed unequal.

The Equal Variances Case

If the population variances are assumed equal ($\sigma_1^2 = \sigma_2^2$), then the test statistic is

$$t^* = \frac{\bar{x}_1 - \bar{x}_2}{\sqrt{\dfrac{(n_1 - 1)\, s_1^2 + (n_2 - 1)\, s_2^2}{n_1 + n_2 - 2}}\sqrt{\dfrac{1}{n_1} + \dfrac{1}{n_2}}}$$

where $t^* \sim t(n_1 + n_2 - 2)$.

In this case the t.test function call can have either of the forms

```
t.test(x, y, alternative, var.equal = TRUE)

t.test(formula, data, alternative, var.equal = TRUE)
```

The argument assignment var.equal = TRUE indicates the population variances are assumed equal, and the remaining arguments serve the same purposes as described for the previous dependent samples case.

The Unequal Variances Case

This test is referred to as *Welch's two-sample t-test*. If the population variances are assumed to be unequal ($\sigma_1^2 \neq \sigma_2^2$), then the test statistic is

$$t^* = \frac{\bar{x}_1 - \bar{x}_2}{\sqrt{s_1^2/n_1 + s_2^2/n_2}}$$

and t^* can be approximated by the t-distribution with degrees of freedom ν obtained using[8]

$$\nu = \frac{\left(s_1^2/n_1 + s_2^2/n_2\right)^2}{\left(s_1^2/n_1\right)^2 \big/ (n_1 - 1) + \left(s_2^2/n_2\right)^2 \big/ (n_2 - 1)}.$$

For this case the t.test function call can be either of

```
t.test(x, y, alternative, var.equal = FALSE)

t.test(formula, data, alternative, var.equal = FALSE)
```

---

[8]See, for example, [12, p. 482]. Some algebra shows that $\nu \geq \min(n_1, n_2) - 1$. The right side of this inequality serves as a conservative lower bound for the degrees of freedom and texts often suggest using this for the degrees of freedom for the approximating t-distribution when working with probability distribution tables.

The argument assignment `var.equal = FALSE` is the default setting, and the remaining arguments serve the same purposes as described previously.

Example

Consider using Doc's dataset from Section 13.3.1 to determine if there is a difference in the true mean weights of Dollies that can be caught at the two locations, that is, test the hypotheses

$$H_0 : \mu_{\mathrm{DB}} = \mu_{\mathrm{PL}} \quad \text{vs.} \quad H_0 : \mu_{\mathrm{DB}} \neq \mu_{\mathrm{PL}}.$$

Based on the result from the example in Section 13.3.2, assume that the variances are equal. Then,

```
> t.test(weight ~ where, data = dolly, var.equal = TRUE)
```

yields $t^* = 0.3207$, $df = 83$, p-value $= 0.7492$ and it may be concluded that for any typical level of significance there is insufficient evidence to conclude that a difference in the true mean weights of Dollies from the two types of locations exists.

If the variances were to be assumed unequal the argument assignment `var.equal = FALSE` would be used. It will be noticed for the example in question that the inference of the hypothesis test is the same.

A Comment about the Confidence Interval Produced: The 95% confidence interval produced by `t.test` for this two-tailed test, to three decimal places, is approximately

$$-0.278 < \mu_{\mathrm{DB}} - \mu_{\mathrm{PL}} < 0.384.$$

Since the hypothesized difference, $\mu_{\mathrm{DB}} - \mu_{\mathrm{PL}} = 0$, lies inside the confidence interval the null hypothesis is not rejected at the $\alpha = 0.05$ level of significance.

15

Tests for More than Two Means

This chapter addresses tests involving more than two population means associated with normally distributed random variables. Topics from Chapter 13 are once again relevant since, in comparing two or more means, the methods applied are determined by whether the population variances are assumed equal or not. Code for all examples, and more, including a collection of functions prepared for this chapter are contained in the script file `Chapter15.Script.R`.

15.1 Relevant Probability Distributions

Probability distributions already seen that find use in this chapter include the standard normal, Student's t and F distributions. Three additional distributions are the studentized range distribution, Dunnett's test distribution and the studentized maximum modulus distribution.

15.1.1 Studentized Range Distribution

Quantiles for this distribution are needed for the multiple comparison procedure introduced by Tukey for balanced experiment designs in [130] and [131], and later modified by Kramer for unbalanced designs in [79].

Let $Y \sim N(\mu, \sigma)$ and let $S = \{y_1, y_2, \ldots, y_p\}$ be a random sample of independent observations from Y. Now let

$$w = \max(S) - \min(S),$$

the range of S, and suppose an independent unbiased estimate, s, of σ based on ν degrees of freedom is available. Then, the sampling distribution of the ratio $X = w/s$ is called the *studentized range distribution*; see [99] for an early introduction to this distribution. Notation such as $X \sim q(p, \nu)$ will be used to indicate that a random variable belongs to a studentized range distribution.

Now, for $j = 1, 2, \ldots, p$ and $i = 1, 2, \ldots, n_j$ denote the mean of the sample $\{y_{1j}, y_{2j}, \ldots, y_{n_j j}\}$ by $\bar{y}_j$. For applications of the studentized range distribution in this chapter, S comprises the set of p such sample means. Let $n = \sum_{j=1}^{p} n_j$, the total number of observations for all p samples. It is assumed that the underlying random variables are normally distributed, that the variances of the p random variables are equal (denoted by σ^2), and that an unbiased estimate s of σ with $\nu = n - p$ degrees of freedom is known.

The two functions available in R for the studentized range distribution are the ptukey and qtukey functions. The function for computing $P(X \leq x)$ is

```
ptukey(q, nmeans, df, lower.tail = TRUE)
```

and for a given probability P, the function

```
qtukey(p, nmeans, df, lower.tail = TRUE)
```

is used to find x such that $P(X \leq x) = P$. The arguments p, q and lower.tail serve the usual purposes; nmeans is assigned p, the number of sample means involved; and df is assigned the degrees of freedom $\nu = n - p$.

There is no analogous dtukey function to compute probability densities. Also, providing a general form for the probability density function for this distribution presents a notational challenge, so it is not given here (see, for example, [99], [104] and [110]). However, the density function does exist and, for the case of $p = 2$, with $x \geq 0$

$$f_q(x) = \sqrt{2}\, f_t\left(x/\sqrt{2}\right)$$

where f_q is the probability density function for $q(2, \nu)$ and f_t is the density function for $t(\nu)$, see [99]. It is possible to use R to demonstrate this relationship graphically by obtaining approximations of probability densities using the function ptukey; see the script file. Figure 15.1 is the result of preparing and taking advantage of a "brute-force" function (named dtukey for consistency) created for the above mentioned demonstration and for this figure.

Examples

Let $X \sim q(5, 93)$, the probability density curve of which is shown in Figure 15.1. Then, calculations using the functions ptukey and qtukey are analogous to the use of similar functions for previously encountered probability distributions. For example, the two function calls

```
> ptukey(q = 3.93, nmeans = 5, df = 93, lower.tail = FALSE)
> qtukey(p = 0.95, nmeans = 5, df = 93, lower.tail = TRUE)
```

compute $P(X > 3.93)$ and find x for which $P(X \leq x) = 0.95$, respectively.

15.1.2   Dunnett's Test Distribution

Dunnett introduced this distribution in [43] and [45] where comparisons of treatment means against a control were considered through a multivariate generalization of the t-distribution. A very brief outline of the development of this distribution is helpful in understanding the arguments used in relevant

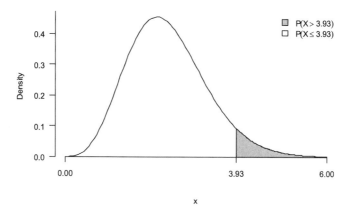

FIGURE 15.1
Density curve of $X \sim q(5, 93)$. The shaded area under the curve represents a right-tailed probability and the unshaded area a left-tailed probability.

R functions. Consider a collection of random variables $X_j \sim N(\mu_j, \sigma)$, $j = 1, 2, \ldots, p$ for which means, denoted by $\bar{x}_j$, are available for representative samples $\{x_{1j}, x_{2j}, \ldots, x_{n_j j}\}$. Let $n = \sum_{j=1}^{p} n_j$ and

$$s^2 = \sum_{j=1}^{p} \sum_{i=1}^{n_j} (x_{ij} - \bar{x}_j) / (n - p).$$

If X_1 represents the random variable associated with the control treatment, for $j \neq 1$, write

$$z_j = \frac{(\bar{x}_j - \bar{x}_1) - (\mu_j - \mu_1)}{\sqrt{1/n_j + 1/n_1}}$$

and denote the correlation between z_j and z_k by ρ_{jk} where

$$\rho_{jk} = 1 \left/ \sqrt{(n_1/n_j + 1)(n_1/n_k + 1)} \right. .$$

Then, for $j = 2, 3, \ldots, p$, the joint distribution of the random variables $t_j = z_j/s$ is the multivariate analog of the t-distribution. The resulting distribution is referred to as *Dunnett's test distribution* with parameters $p-1$ and ν, where p is the number of treatments (including the control) and $\nu = n-p$ the degrees of freedom. Notation to indicate such a distribution will be $d(p - 1, \nu)$.

Of interest in applications of this distribution are critical values, denoted by d_α, for which

$$P(|t_2| < d_\alpha \ \& \ |t_3| < d_\alpha \ \& \ \cdots \ \& \ |t_p| < d_\alpha) = 1 - \alpha$$

As with the studentized range distribution, notational challenges arise in describing the relevant density and distribution functions. However, R functions for these are available in package nCDunnett [17]. The two functions that are of interest are

```
qNCDun(p, nu, rho, delta, n = 32, two.sided = TRUE)

pNCDun(q, nu, rho, delta, n = 32, two.sided = TRUE)
```

The arguments p and q both can be assigned a single value or a vector of values and serve the same purposes as in previous quantile and probability functions. The argument two.sided is used to indicate whether a two-sided versus a one-sided probability should be used in computations. As for the remaining arguments: nu represents the degrees of freedom; rho is assigned a vector of correlations, having the same number of entries as there are treatments (excluding the control); and delta a vector with the same length as rho, this is the non-centrality parameter and for applications in this book this will always be a vector of zeros. The argument n is a positive integer used to determine the accuracy of the computations; the higher the value of n, the greater the degree of accuracy. Note that the larger the value assigned to n the longer it takes for computations to be performed, particularly for qNCDun.

The vector assigned to rho has as its entries the correlations

$$\rho_{jj} = n_j \, / (n_1 + n_j)$$

where $j = 2, 3, \ldots, p$. If the experiment design is balanced, that is, if $n_1 = n_2 = \cdots = n_p$, then the correlations all equal $1/2$. Tables of Dunnett's test distribution constructed by Dunnett in [43] and reproduced elsewhere use samples of equal sizes, and the default setting of n = 32 in the function qNCDun produces quantiles to the same decimal place accuracy available in these tables.

Examples

Consider the case of a balanced experiment design having 4 treatments and a control with samples of size 5 each, then the Dunnett test distribution in question is $d(4, 20)$. Begin by loading the package nCDunnett into the workspace. Since $n_1 = n_2 = \cdots = n_5 = 5$, $\rho_{jj} = 0.5$ for $j = 2, 3, \ldots, 5$, and the function call

```
> qNCDun(p = 0.95, nu = 20, n = 50, two.sided = TRUE,
+       rho = c(0.5, 0.5, 0.5, 0.5), delta = c(0, 0, 0, 0))
```

computes the positive critical value for a 95% two-sided confidence interval associated with this distribution. Alternatively,

```
> pNCDun(q = 2.651227, nu = 20, n = 50, two.sided = TRUE,
+       rho = c(0.5, 0.5, 0.5, 0.5), delta = c(0, 0, 0, 0))
```

estimates the two-sided probability for the given quantile associated with this distribution.

15.1.3 Studentized Maximum Modulus Distribution

Quantiles for this distribution are needed for the T3 multiple comparison procedure introduced by Dunnett in [44]. A brief description of this distribution appears in [127] and is presented here somewhat along the lines of the introduction to the studentized range distribution in Section 15.1.1.

Let $Y \sim N(\mu, \sigma)$ and $S = \{y_1, y_2, \ldots, y_k\}$ a random sample of independent observations from Y. Now let

$$z_{ij} = y_i - y_j, \quad \text{where } 1 \le i < j \le k,$$

then the resulting sample of differences, having size $r = k(k-1)/2$, belongs to $Z \sim N(0, \sigma)$. Next, let s^2 be an estimate of σ^2 which is distributed as $\sigma^2 \chi^2(\nu)/\nu$, where ν represents the degees of freedom of the associated chi-square distribution. Then, the sampling distribution of the statistic

$$m = \max_{1 \le i < j \le k} |z_{ij}| / s$$

is called the *studentized maximum modulus distribution* with parameters r and ν. Notation of the form $M \sim M(r, \nu)$ will be used to identify a random variable as belonging to such a distribution.

Of interest, for Dunnett's T3 procedure, are upper-tail quantiles m_α for which $P(M \ge m_\alpha) = \alpha$. Tables of such quantiles computed and provided in [127] can be duplicated using the function

```
qsmm(p, k, nu, lower.tail = TRUE)
```

Alternatively, probabilities corresponding to a given quantile can be obtained using the function

```
psmm(q, k, nu, lower.tail = TRUE)
```

In these functions[1] the arguments p and q, and the default argument assignment lower.tail = TRUE holds the same meaning as for previously encountered quantile and probability functions. The arguments k and nu are as defined in the preceeding discussion. For applications of these functions in this book the argument k is assigned the number of means being compared.

---

[1]Code for these functions were adapted from the function rgabriel, contained in package rgabriel [143], which implements the multiple comparison method described in [57].

Examples

Consider the case for $k = 7$ with $\nu = 24$. Since $r = k(k-1)/2 = 21$, the distribution in question is $M \sim M(21, 24)$. To find the quantile m corresponding to the right-tailed probability $P(M > m) = 0.01$, run

```
> qsmm(p = 0.01, k = 7, nu = 24, lower.tail = FALSE)
```

and, to calculate the probability $P(M \leq 4.019601)$, run

```
> psmm(q = 4.019601, k = 7, nu = 24, lower.tail = TRUE)
```

15.2   Setting the Stage

Consider a study in which independent random samples are obtained from $p > 2$ populations. Denote the underlying random variables by $Y_1, Y_2, \ldots, Y_p$ and denote the corresponding means by $\mu_1, \mu_2, \ldots, \mu_p$, respectively. For $j = 1, 2, \ldots, p$ and for $i = 1, 2, \ldots, n_j$, let y_{ij} denote the ith observed response from the jth population. Here, n_j denotes the sample size from the jth population and the p sample sizes need not be all equal. An experiment in which the sample sizes are the same, that is, $n_1 = n_2 = \cdots = n_p$, is said to have a *balanced design*. Otherwise, the design is said to be *unbalanced*.

Matters of Interest

There are two questions that one might ask concerning the p population means. First, is there sufficient evidence to conclude that the population means are not all equal?

The answer to this question is obtained by testing the hypotheses

$$H_0 : \mu_1 = \mu_2 = \cdots = \mu_p \quad \text{vs.} \quad H_1 : \text{The means are not all equal.}$$

An alternative view of this problem is to consider a single random response variable Y as being dependent on levels of a factor representing treatments on subjects in a designed experiment. From this point of view testing the above hypotheses is equivalent to assessing the effects of varying treatments on the response variable Y.

If statistically significant differences between the population means are found to exist, the nature of the differences may be of interest. In this case a set of *simultaneous pairwise comparisons* of treatment means is conducted through tests of hypotheses having the form

$$H_0 : \mu_j = \mu_k \quad \text{vs.} \quad H_1 : \mu_j \neq \mu_k,$$

for pairs $j, k = 1, 2, \ldots, p$ with $j \neq k$.

It is worth mentioning that the alternative hypothesis of the global test actually covers a wide range of possibilities, so in some cases it may be of interest to perform a test on a more *general linear contrast* along the lines of that described for proportions in Section 12.3. General linear contrasts involving means are not covered in this book. However, the function `glht` in package `multcomp` addresses these and much more. See [16], and the user guides and other documentation available for package `multcomp` [71] for details and examples.

The Assumptions

It is assumed that the populations (hence samples) are independent, and the observed responses, y_{ij}, are independent of each other within each sample.

As with previously covered tests of means, it is assumed that the underlying random variables $Y_1, Y_2, \ldots, Y_p$ are (at least approximately) normally distributed. If the underlying random variables are normally distributed, then methods described in this chapter are applicable. However, something needs to be known about the population variances. When the p population variances $\sigma_1^2, \sigma_2^2, \ldots, \sigma_p^2$, are assumed equal the methods described in Sections 15.3 and 15.4 are applicable, and when they are not the methods described in Sections 15.5 and 15.6 may be used.

These assumptions can be equivalently restated. For $j = 1, 2, \ldots, p$, denote the sample mean for the jth sample by $\bar{y}_j$. Then for $i = 1, 2, \ldots, n_j$ the quantities

$$\hat{\varepsilon}_{ij} = y_{ij} - \bar{y}_j$$

are referred to as *residuals* and the corresponding underlying *random error terms/variables*, ε_{ij}, are assumed independently distributed with $\varepsilon_{ij} \sim N(0, \sigma_j)$, where the σ_j^2 are the unknown and not necessarily equal population variances.

Verifying the Assumptions

If proper randomization is not achieved in an experiment, a violation of the independence assumption occurs and the introduced bias affects any tests or estimations performed. For all that follows in this chapter it is assumed that an appropriate experiment design is employed and that the underlying random variables and observed responses satisfy the independence assumption.[2]

In assessing the assumptions of normality and constant variances associated with the methods under discussion, the common practice is to use the residuals to assess the equivalent assumptions on the error terms. This being said, a scaled version of the residuals can be used for the same purpose. Let

---

[2]See, for example, [80, Ch. 15] for a detailed discussion on the design of experimental and observational studies.

$n = \sum_{j=1}^{p} n_j$, then the *mean-square error* is defined by

$$s^2 = \sum_{j=1}^{p} \sum_{i=1}^{n_j} \hat{\varepsilon}_{ij}^2 \bigg/ (n - p),$$

and the *residual standard error* s can be used to scale (standardize) these residuals through $\hat{r}_{ij} = \hat{\varepsilon}_{ij}/s$. The resulting *standardized residuals* can then be used in the graphical analyses of the residuals. See, for example, [65, pp. 173–174 and Sec. 11.4] for more on this topic and related scalings with respect to R and as it applies to such analyses.

The validity of the normality assumption on the error terms and hence the underlying response variable can be assessed graphically by means of a normal probability QQ-plot of the residuals (or standardized residuals) and, numerically, by means of the Shapiro–Wilk test.

The validity of the constant variances assumption may be assessed by means of stripcharts and/or error bars of the residuals (or standardized residuals) by treatments. These provide an effective visual method of diagnosing large differences in variances. For marginal differences one of the tests discussed in Section 13.4 can be used to help make a decision.

What If the Normality Assumption Is Violated?

Violations of the normality assumption affect the p-value obtained for the commonly used one-factor fixed-effects ANOVA F-test described in Section 15.3 in the following ways.

First the good news. Small to moderate violations of the normality assumption can typically be ignored since the effect on the p-value is small. This applies to both small and large sample scenarios, and balanced designs help the cause. Next, if the design is balanced, long but symmetric tails make the F-test conservative (less likely to reject the null hypothesis) and short but symmetric tails make the test liberal (more likely to reject the null hypothesis). Also, assymmetry has less of an effect on the p-value than tail length. Finally, for unbalanced designs, the effects of non-normality vary from unpredictable for small samples to negligible for large samples.

A violation of the normality assumption can sometimes be remedied through a transformation of the response variable. However, in doing this the resulting test at best becomes equivalent to performing a test on the medians associated with the (original) untransformed response variable. For this reason one might consider performing tests directly on medians. The next chapter addresses commonly used tests for such purposes.

Implications of Unequal Variances

In general, a violation of the constant variances assumption affects the power of the one-way ANOVA F-test, and the extent to which this happens depends on whether the experiment design is balanced or not.

A benefit to having balanced designs is that the effect of unequal variances across treatments on the p-value is small; the violation can be ignored. This is particularly the case when the largest sample variance is less than or equal to three times the smallest sample variance. See, for example, [37, p. 112].[3]

If the design is unbalanced and larger variances correspond to larger sample sizes the result is a conservative test. On the other hand, if smaller variances correspond to larger sample sizes the result is a liberal test.

As with deviations from the normality assumption, transformations of the response variable can be used to remedy violations of the constant variances assumption. Discussions of this and other remedies can be found in, for example, [47, Ch. 6 and 7] and [65, Ch. 15]. Alternatively, methods developed to combat violations of the constant variances assumption can be used, some of which are presented in Section 15.5.

15.3   Equality of Means: Equal Variances Case

For the method discussed in this section it is assumed that the error terms, ε_{ij}, are independently and identically distributed with $\varepsilon_{ij} \sim N(0, \sigma)$ where σ^2 represents the (unknown and common) population variance.

For this case a *one-factor fixed-effects analysis of variance*, more commonly called the *one-way ANOVA*, is used to answer the first question raised in Section 15.2. That is, to test the hypotheses

$$H_0 : \mu_1 = \mu_2 = \cdots = \mu_p \quad \text{vs.} \quad H_1 : \text{ The means are not all equal.}$$

Let p, n_j, n and $\bar{y}_j$ be as previously defined, denote the overall mean of (all) the observed responses by $\bar{y}$, and the jth sample variance by s_j^2. Then the test statistic use to test the equality of the p means (or treatment effects) is[4]

$$F^* = \frac{\sum_{j=1}^{p} n_j \left(\bar{y}_j - \bar{y} \right)^2 \Big/ (p - 1)}{\sum_{j=1}^{p} (n_j - 1) s_j^2 \Big/ (n - p)} \sim F(p - 1, n - p)$$

and p-value $= P(F \geq F^*)$.

If the response vector is denoted by `response` and the factor by `treatment`, then the above calculations can be performed and the results outputted using code of the form:

---

[3]Some suggest a higher threshold. For example, in [55, pp. 276–277] it is suggested that the assumption of equal variances is probably not violated if the ratio of the largest sample variance to the smallest sample variance does not exceed 4.

[4]The numerator in this test statistic is an estimate of the mean-square of variations due to the treatments, and the denominator is an estimate of the mean-square of variations due to random fluctuations, or pure error.

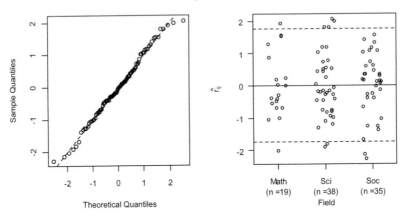

FIGURE 15.2
QQ normal probability plot (left) and stripchart (right) of the standardized residuals. Note, the dashed lines in the stripchart are for eye-balling/reference purposes only.

```
model <- lm(formula = response ~ treatment, data, subset)
anova(model)
```

The theory behind the methods that the function `lm` employs is beyond the scope of this book.[5] However, the results are identical to using the test statistic formula given above and it is convenient to take advantage of the simplicity of using the `lm` and `anova` functions.

Example

Recall the example in Section 13.4. There, a test of the equality of the variances of scores in the `elemStatStudy` data frame across fields, and within the traditional course only, yielded insufficient evidence to reject an assumption of equal variances.

For this example, consider testing the hypotheses

$$H_0 : \mu_{\text{Math}} = \mu_{\text{Sci}} = \mu_{\text{Soc}} \quad \text{vs.} \quad H_1 : \text{The means are not all equal,}$$

---

[5] One may also use

```
summary(aov(formula = resp ~ treatment, data))
```

The `aov` function is referred to as an `lm` (linear model) *wrapper*. For the one-way ANOVA, combined with the ever-useful `summary` function, the `aov` function makes use of the `lm` function to output the same results as the above code.

once again within the traditional course only. First fit the data to the model and extract the standardized residuals by running

```
> model <- lm(formula = score ~ field,
      data = elemStatStudy, subset = (course == "trad"))
> modelSummary <- summary(model)
> r <- residuals(model)/modelSummary$sigma
```

The quantity `sigma`, contained in the object `modelSummary` is the residual standard error. Then, construct residual plots (Figure 15.2) to graphically verify the normality assumption and constant variances assumption on the residuals.

A quick verification of the normality and constant variances assumptions can be accomplished through the two plots in Figure 15.2. The normal probability QQ-plot does not suggest a violation of the normality assumption, and the stripchart does not (strongly) suggest a difference in variances across fields. Moreover, formal tests do not suggest a contradiction of these observations (see code in the script for this section).

So, even though the design is unbalanced, there are no concerns with respect to continuing with testing the hypotheses. To wrap things up, run

```
> anova(model)
```

to get $F^* = 0.6711$, $\nu_1 = 2$, $\nu_2 = 89$ and p-value $= 0.5137$.[6] So, at traditional levels of significance, there is insufficient evidence to conclude that statistically significant differences between means across fields for the traditional course exist.

15.4   Pairwise Comparisons: Equal Variances

Recall that the answer to the second question posed in Section 15.2, if applicable, is obtained by performing a set of *simultaneous pairwise comparisons* of treatment means through tests of hypotheses having the form

$$H_0 : \mu_j = \mu_k \quad \text{vs.} \quad H_1 : \mu_j \neq \mu_k,$$

for each pair $j, k = 1, 2, \ldots, p$ with $j \neq k$. Observe that if one were to wish to test all possible pairwise differences, then there will be $m = p(p-1)/2$ pairwise

---

[6]Identical, but differently formated results can be obtained using the function `oneway.test` in a call of the form

```
> oneway.test(formula = score ~ field, var.equal = TRUE,
+     data = elemStatStudy, subset = (course == "trad"))
```

comparisons in all, and if the matter of interest is to compare $p - 1$ of the treatments against a control, there will be $m = p - 1$ pairwise comparisons.

Two popular methods for the equal variances case addressing all possible pairwise comparisons are Bonferroni's and Tukey's procedures. For comparisons with a control, while Bonferroni's procedure may be used, Dunnett's test is a method designed specifically for the purpose.

15.4.1   Bonferroni's Procedure

The assumptions for this procedure are the same as for the one-way ANOVA. That is, the underlying random variables are independent and normally distributed with equal variances. The formula for the $m = p(p - 1)/2$ simultaneous $100(1 - \alpha)\%$ Bonferroni confidence intervals for the difference between two treatment means is

$$(\bar{y}_j - \bar{y}_k) - t_{\alpha/(2m)}\, s\, \sqrt{\frac{1}{n_j} + \frac{1}{n_k}} < \mu_j - \mu_k < (\bar{y}_j - \bar{y}_k) + t_{\alpha/(2m)}\, s\, \sqrt{\frac{1}{n_j} + \frac{1}{n_k}},$$

where s is the previously defined standard error and $t_{\alpha/(2m)}$ is such that $P\left(t \geq t_{\alpha/(2m)}\right) = \alpha/(2m)$ with $t \sim t(n - p)$.

Two population means, say μ_j and μ_k, are considered different at the *individual* α/m level of significance if the associated confidence interval for $\mu_j - \mu_k$ does not contain zero.

The corresponding test statistic is

$$t_{jk}^* = \frac{\bar{y}_j - \bar{y}_k}{s\,\sqrt{1/n_j + 1/n_k}} \sim t(n - p),$$

and, being a two-sided test, p-value $= 2P\left(t \geq \left|t_{jk}^*\right|\right)$ for each comparison. Then, each of the m simultaneous tests are performed by comparing each p-value against the individual significance level α/m.[7]

The `pairwise.t.test` function performs the above described computations, the basic usage definition with arguments of interest being

```
pairwise.t.test(x, g, p.adjust = "bonferroni")
```

where `x` is assigned the vector containing the observed responses and `g` is assigned the factor containing the treatment levels that identify the samples.

The adjusted p-values outputted with the `p.adjust = "bonferroni"` option are of the form $m \times$p-value,[8] the `alternative` argument is left in its

---

[7]Alternatively, each $m \times$p-value can be compared against the joint significance level α. In cases where $m \times$p-value > 1 assign an adjusted p-value of 1.

[8]If `p.adjust = "none"` is assigned, the results of one-at-a-time *Least Significant Difference* tests are outputted. In the statistically legal sense, only a single pairwise comparison is permitted in this case since the individual significance level for each pairwise comparison is α, and the computed p-values are not multiplied by m before being outputted.

default setting, and this function does not output the corresponding simultaneous confidence intervals.

Example

Using the data in `elemStatStudy`, consider looking at the mean scores by fields for the project-based scores only, and at the $\alpha = 0.10$ level of significance. The hypotheses are

$$H_0 : \mu_{\text{Math}} = \mu_{\text{Sci}} = \mu_{\text{Soc}} \quad \text{vs.} \quad H_1 : \text{The means are not all equal,}$$

Following the track of the previously described one-way ANOVA F-test (see script file for this part), analyses of the residuals do not provide evidence of serious violations of the relevant assumptions of normality and constant variances.

On conducting the one-way ANOVA, it is found that there is sufficient evidence to reject the null hypothesis ($F^* = 2.8741$, $\nu_1 = 2$, $\nu_2 = 80$, p-value $= 0.06232$). So, one might expect at least one mean score to differ from the others.

Now, to determine the nature of the differences, run

```
> with(data = subset(x = elemStatStudy,
+           subset = (course == "proj")),
+       expr = pairwise.t.test(x = score, g = field,
+           p.adjust = "bonferroni"))
```

to get the p-values (in the form of a matrix). It will be noticed that the adjusted p-value for the difference $\mu_{\text{Sci}} - \mu_{\text{Soc}}$ is the only one that is less than $\alpha = 0.10$, so this is the only statistically significant difference.

Side Comments: In the preliminary analyses for this example one might have expected the mean scores of the mathematical sciences and social sciences to also differ significantly. The fact that this difference is not found to be statistically significant at $\alpha = 0.10$ might be surprising. However (see code in the script file), it can be observed that the size of the mathematical sciences sample is about half that of the social sciences sample, and the sample mean for the mathematical sciences scores is slightly lower than that of the natural sciences, but not by much.

Taking this information to the formula for the confidence interval suggests a quantitative explanation of the results. For given sample means and residual standard errors, larger differences in sample sizes produce wider confidence intervals resulting in a test that is less likely to justify a rejection of a null hypothesis. From the statistical point of view, this means that unequal sample sizes can have the effect of reducing the power of the test; the larger the difference in sample sizes, the greater the drop in power. See the script file for a simulation using the independent samples t-test with equal variances and with ratios of sample sizes ranging over $0.1 \leq n_1/n_2 \leq 1$ to demonstrate this effect.

15.4.2　Tukey's Procedure

For any distinct pair $j, k = 1, 2, \ldots, p$ for which $\bar{y}_j \geq \bar{y}_k$, the formula for all possible $100(1 - \alpha)\%$ simultaneous Tukey confidence intervals for the differences $\mu_j - \mu_k$ is defined as follows

$$(\bar{y}_j - \bar{y}_k) - \frac{1}{\sqrt{2}} q_\alpha \, s \sqrt{\frac{1}{n_j} + \frac{1}{n_k}} < \mu_j - \mu_k < (\bar{y}_j - \bar{y}_k) + \frac{1}{\sqrt{2}} q_\alpha \, s \sqrt{\frac{1}{n_j} + \frac{1}{n_k}},$$

where s is the same as for the previous section.[9] The critical value $q_\alpha > 0$ is such that $P(q \geq q_\alpha) = \alpha$ with $q \sim q(p, n - p)$.

Alternatively, the corresponding test statistic

$$q^*_{jk} = \frac{\sqrt{2}(\bar{y}_j - \bar{y}_k)}{s\sqrt{1/n_j + 1/n_k}} \sim q(p, n - p)$$

can be used to calculate p-value $= P(q \geq q^*_{jk})$. For Tukey's procedure, by design the individual and joint significance levels are the same so the p-value for each pairwise difference is compared against α.

The R function to implement Tukey's procedure for balanced or mildly unbalanced designs is

```
TukeyHSD(x, which, ordered = FALSE, conf.level = 0.95)
```

The argument x is assigned an aov object, which is assigned a vector of characters that identify which means are to be included in the pairwise comparisons. If none are assigned, this defaults to all pairwise comparisons. If ordered = TRUE is used the pairwise differences in means are computed after sample means are arranged in ascending order and, under this setting, statistically significant differences in means exist for those pairs that yield a positive lower interval bound.

Example

To apply the function TukeyHSD to the previous example, run

```
> with(data = subset(x = elemStatStudy,
+            subset = (course == "proj")),
+        expr = TukeyHSD(x = aov(formula = score ~ field),
+            ordered = TRUE, conf.level = 0.90))
```

and the ever-useful plot function can be used to plot the confidence intervals; see this section's script. It will be observed that the inferences from this procedure are similar to those from Bonferroni's procedure.

---

[9]The name *Tukey–Kramer procedure* is sometimes given to this procedure when it is applied to samples of unequal sizes.

15.4.3 t Tests and Comparisons with a Control

For the sake of convenience, consider the first treatment as always representing the control; the lm function (and the aov function) uses the first treatment as the base-line treatment. As such, what is involved here are tests of hypotheses having the form

$$H_0 : \mu_j = \mu_1 \quad \text{vs.} \quad H_1 : \mu_j \neq \mu_1,$$

for each $j = 2, 3, \ldots, p$.

The approach and computations described here are as described for Bonferroni's procedure in Section 15.4.1, except that here the number of pairwise comparisons is $m = p - 1$, the number of treatments (excluding the control). There are two ways in which the needed computations can be performed.

Computations for comparison of a collection of treatments against the baseline treatment are built into the summary (or summary.lm) function when applied to the fitted model object as described in Section 15.3 and outputted under the heading coefficients. The hypotheses for these comparisons are equivalent to the hypotheses in question here. For example, see [65, Sec. 11.3 and 11.5] for a brief explanation of this equivalence.

For the present, what is relevant is that if a collection of observed responses are classified by levels contained in a factor as described in Section 15.2 and passed into the function lm (or aov) as described in Section 15.3 then, by default, the first level in the factor is used as the baseline treatment in the computations performed by the lm (or aov) function. So, if the control treatment is the first level in the factor, then the second line onward in the coefficients component of the output produced by the summary (or summary.lm) function contains the estimated differences between the mean responses of each treatment and the control, the associated t-statistics, and the corresponding two-tailed p-values. By applying the Bonferroni adjustment to a joint significance level of α, the resulting p-values can then be used to draw inferences about differences between each of the treatment means and the control by comparing the p-values against the adjusted significance levels $\alpha/(p-1)$. As illustrated in the following example, the pairwise.t.test function can be used for the same purpose.

Example

Consider a lawn-care illustration involving a common complaint of homeowners in Juneau, the matter of moss in lawns.[10] Every Spring, exasperated owners of homes with lawns consider ways by which they can reduce the amount of moss in their lawns. Typically, a thatcher is used to remove as much moss as possible on the first clear day after the snow has melted and then various chemical moss-treatments are available to keep the moss down until the snow arrives in the Fall. Another approach is to simply torch the

---

[10]Being located within the boundaries of the largest temperate rainforest in the world, moss grows almost everywhere and on almost anything in Juneau.

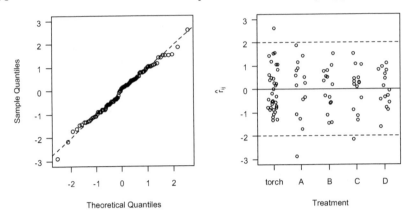

FIGURE 15.3

Residual plots for the model associated with the analysis of the moss data. The normal probability QQ-plot (left) does not provide strong evidence of a violation of the normality assumption, and the stripchart (right) does the same with regard to the constant variances assumption. The Shapiro–Wilk and Levene's tests do not contradict these observations.

whole lawn, again on the first clear day, and leave it at that. The question at hand is the following: Do thatcher-chemical treatments work better at discouraging moss than the torching technique? Lonnie Green, an advocate for a chemical-free environment, decided to conduct a study to find the answer by comparing the effects of four moss-treatment chemicals against the torching technique. Ninety five equivalent lawns were selected for the study, and from these 15 randomly selected lawns were assigned to each treatment (brands A, B, C and D) and 35 to the control (torching). Then, during the first sunny week in April each yard was prepared either through torching, or through thatching and then a chemical treatment at equal intervals for five months. Exactly five months later the amount of moss coverage on each lawn was measured as a percentage of lawn area. The data are contained in the data frame moss under percent and treatment.

To compare the effects of each thatcher-chemical treatment against the control, begin the process by fitting the data to the one-factor linear model using the lm function as described in Section 15.3. That is, run

```
> model <- lm(formula = percent ~ treatment, data = moss)
```

Next, verify the normality and constant variances assumptions on the error terms. This can be done graphically, see Figure 15.3, as well as numerically if so desired; see the script file for the code used. Then, from

```
> anova(model)
```

it is evident that, at $\alpha = 0.05$, there exist statistically significant differences in mean moss coverage among the treatments applied ($F^* = 3.341$, $\nu_1 = 4$, $\nu_2 = 90$ and p-value $= 0.0343$). Next, to determine how the treatment means compare against the control mean, run

```
> modelSummary$coefficients
```

Identical p-values can be obtained from the first column of output from the `pairwise.t.test` function, but without activating the Bonferroni adjustment (see the script file). Either way, the results are the same and can be viewed as follows.

Null Hypothesis	Difference Estimate	p-value
$H_0 : \mu_A - \mu_{Control} = 0$	1.99	0.0296
$H_0 : \mu_B - \mu_{Control} = 0$	1.21	0.1825
$H_0 : \mu_C - \mu_{Control} = 0$	−1.54	0.0891
$H_0 : \mu_D - \mu_{Control} = 0$	−0.28	0.7589

To complete the process, the p-values are compared against $\alpha/4 = 0.0125$.[11]

So, according to this test, and in the statistically significant sense, none of the treatments have an effect on moss growth that differs from the effect of the control treatment.

15.4.4   Dunnett's Test and Comparisons with a Control

This test, introduced in [43], is considered to be the most powerful test for purposes of comparing multiple treatments against a control. As before, let μ_1 denote the true mean under the control treatment. Then, with $j = 1$ being reserved for the control treatment, for any $j = 2, \ldots, p$ the formula for the $p-1$ simultaneous $100(1-\alpha)\%$ Dunnett confidence intervals for the differences $\mu_j - \mu_1$ is

$$(\bar{y}_j - \bar{y}_1) - d_\alpha s \sqrt{\frac{1}{n_j} + \frac{1}{n_1}} < \mu_j - \mu_1 < (\bar{y}_j - \bar{y}_1) + d_\alpha s \sqrt{\frac{1}{n_j} + \frac{1}{n_1}},$$

where s is the standard error as previously defined. The critical value $d_\alpha > 0$ is as defined and computed in Section 15.1.2 for Dunnett's test distribution

---

[11] Alternatively, one can implement the appropriate Bonferroni adjustment on the `pValues` object by running

```
> round(ifelse(test = pValues[,1]*4 < 1, yes = pValues[,1]*4, no = 1),
4)
     A      B      C      D
0.1186 0.7299 0.3592 1.0000
```

and then comparing these adjusted p-values against the joint significance level of $\alpha = 0.05$.

with parameters $p-1$ and $n-p$. Alternatively, the corresponding test statistic

$$d_j^* = \frac{\bar{y}_j - \bar{y}_1}{s\sqrt{\dfrac{1}{n_j} + \dfrac{1}{n_1}}}$$

can be used to calculate the p-value in the manner described in Section 15.1.2.

It is worth mentioning a comment in [94, p. 108] and [43] about sample sizes when dealing with comparisons against a control. It is suggested that if $p-1$ treatments all having samples of the same size, say a, are compared with a control treatment, then the size of the control sample should be chosen to be approximately $a\sqrt{p}$ to maximize the power of the test.

The function `dunnett.test` performs the necessary calculations,[12] its usage definition being

```
dunnett.test(y, g, alpha = 0.05)
```

where, as before the argument `y` is assigned the vector of observed responses, `g` the corresponding grouping factor, and `alpha` the desired level of significance.

Example

Revisit the moss example using Dunnett's test with the help of the function `dunnett.test`. If not already done, first load package `nCDunnett` into the workspace and then run (output not shown)

```
> with(data = moss,
+       expr = dunnett.test(y = percent, g = treatment))
```

Here too none of the treatments are shown to have an effect on moss growth that differs from the effect of the control treatment. However, it will be observed that in all four cases the corresponding Bonferroni adjusted 95% intervals are wider than the Dunnett intervals.

15.4.5 Which Procedure to Choose

The two popular methods for all possible pairwise comparisons are the Bonferroni and Tukey procedures. The interval formulas for these two methods differ only in what the term $s\sqrt{1/n_j + 1/n_k}$ is multiplied by to give the maximum error of estimate for the method in question: $t_{\alpha/(2m)}$ for Bonferroni's procedure and $q_\alpha/\sqrt{2}$ for Tukey's procedure. For a given number of means, p, and combined sample size n the multiplier $q_\alpha/\sqrt{2}$ remains the same irrespective of the number of pairwise comparisons, m, that are to be performed. On the other hand, the magnitude of $t_{\alpha/(2m)}$ depends on m: the higher the number of pairwise comparisons to be performed, the larger the magnitude of $t_{\alpha/(2m)}$

---

[12]Package `nCDunnett` will need to be loaded into the workspace to run `dunnett.test`.

and the result is a wider interval. So, Bonferroni's procedure becomes more conservative as the number of pairwise comparisons intended increases.

Prior to performing pairwise comparisons, it is suggested in [80, p. 757] that a numeric comparison of the two mulipliers be performed to determine which method yields the narrowest confidence intervals for the desired number of pairwise comparisons, given p and n. For example, if (average) sample sizes of 10 are used with $\alpha = 0.05$, the maximum number of pairwise comparisons, m, involving p means that ensures $t_{\alpha/(2m)} < q_\alpha/\sqrt{2}$ are shown in Table 15.1. Another point to note is that since the width of Bonferroni intervals depends on the number of pairwise comparisons to be performed, this method does not lend itself to *data-snooping*. This handicap does not, however, apply to Tukey's procedure.

For comparisons of two or more treatments with a control treatment, Dunnett's procedure is preferable over *both* the Bonferroni and Tukey methods.

TABLE 15.1

The maximum number of pairwise comparisons, m, for which the Bonferroni's procedure would be preferred over Tukey's method. The quantity $p(p-1)/2$ represents the number of all possible pairwise comparisons.

n	30	40	50	60	70	80	90	100	110
p	3	4	5	6	7	8	9	10	11
$p(p-1)/2$	3	6	10	15	21	28	36	45	55
max m	2	4	7	10	14	19	24	30	36

15.5 Equality of Means: Unequal Variances Case

As mentioned in Section 15.2, if the ratio of the largest sample variance to the smallest sample variance is over 3 it is suggested that the usual F-test should not be employed. Here, three alternative methods that address the unequal variances case are described: A large sample chi-square test; Welch's test for small samples; and Hotelling's T^2 test for balanced designs.

15.5.1 Large-Sample Chi-Square Test

This method is described in [87, Ex. 4] as a large-sample method based on a chi-square analog of Sheffe's method described in [118, Ch. 2 and Sec. 3.5].[13]

---

[13]In [87, Ex. 4] samples of size 32 are used. There it is also mentioned that a correction, described in [82, pp. 436–438], can be employed in applications of this method to small samples. The correction alluded to actually amounts to Welch's test, which is described in Section 15.5.2.

For identification purposes, this method will be referred to as *Marascuilo's equality of means test*.

Let p, n_j, $\bar{y}_j$ and s_j^2 be as defined previously and, under the null hypothesis, denote the unknown common mean for the p treatments by μ_0 and the estimate this parameter by $\bar{y}_0$, where

$$\bar{y}_0 = \sum_{j=1}^{p} \frac{n_j \bar{y}_j}{s_j^2} \left/ \sum_{j=1}^{p} \frac{n_j}{s_j^2} \right. .$$

If the error terms are independently distributed with $\varepsilon_{ij} \sim N(0, \sigma_j)$, where the σ_j^2 are the (unknown and not necessarily equal) true variances under each treatment, then for sufficiently large samples

$$\chi^{2*} = \sum_{j=1}^{p} \frac{n_j (\bar{y}_j - \bar{y}_0)^2}{s_j^2} \sim \chi^2(p-1)$$

is the test statistic used in testing the equality of the p treatment means, and p-value $= P(\chi^2 \geq \chi^{2*})$.

The function `marascuiloMeans.test` performs the necessary computations for this test, its usage definition being

```
marascuiloMeans.test(y, g)
```

where the arguments y and **g** are assigned the vector containing the observed responses and the corresponding grouping factor, respectively.

Example

Consider the scenario where a college instructor is interested in comparing scores obtained by students from different age groups on their first exam in a GER mathematics course they are taking for the first time. The data frame `algebraScores`, containing the vector `score` and the factor `age`, is used for this illustration. The age groups considered, in years, are: below 18, $[18, 22)$, $[22, 26)$, $[26, 30)$, and 30 or higher; and the test scores are expressed as percentages to one decimal place. It is safe to assume that the age groups are independent of each other, assume also that the observed scores are independent, both within and across age groups.

An examination of the test scores does not provide strong evidence to suggest that the underlying random variables within each age group deviate from normality (see code in the script file). However, the stripchart and error bars in Figure 15.4 do suggest very strongly that the variances of the scores across age groups are not equal. Levene's test provides evidence to support this conjecture. Moreover, the sample sizes are not equal so the design is not balanced, and larger spreads (variances) correspond to larger sample sizes. Consequently, if the previously described F-test were to be used, the result would be a conservative test.

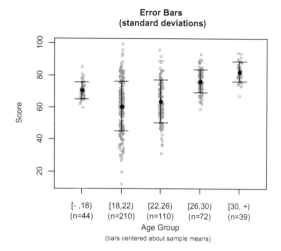

FIGURE 15.4

Stripchart and error bars of test scores for each age group, with scores being centered about sample means.

Denote the true mean test scores for the five age groups (in order) by μ_1, μ_2, μ_3, μ_4 and μ_5, then the hypotheses to be tested are

$$H_0 : \mu_1 = \mu_2 = \mu_3 = \mu_4 = \mu_5 \quad \text{vs} \quad H_1 : \text{The means are not all equal.}$$

To perform Marascuilo's equality of means test run

```
> with(data = algebraScores,
+      expr = marascuiloMeans.test(y = score, g = age))
```

to get $\chi^{2*} = 300.982$, $\nu = 4$ and p-value $= 0$. So, at traditional levels of significance, there is sufficient evidence to conclude that at least one statistically significant difference between mean scores across age groups exists.

15.5.2 Welch's F Test

This test, introduced in [138], is suitable for small samples when $p \leq 4$, [123, p. 550]. For this procedure the test statistic is computed using much of the notation from Section 15.5.1. Let n_j, $\bar{y}_j$, and s_j^2 be as defined in the previous section and $\bar{y}$ the overall mean response. Then the test statistic is

$$F^* = \sum_{j=1}^{p} \frac{n_j \left(\bar{y}_j - \bar{y}\right)^2}{(p-1)\, s_j^2} \bigg/ \left(1 + \frac{2\,(p-2)}{3\nu_2}\right) \sim F(\nu_1, \nu_2),$$

where the degrees of freedom of the numerator are $\nu_1 = p - 1$ and the degrees of freedom of the denominator are estimated by

$$\nu_2 = \cfrac{1}{\cfrac{3}{p^2 - 1} \sum_{j=1}^{p} \cfrac{1}{n_j - 1} \left(1 - \cfrac{n_j / s_j^2}{\sum_{i=1}^{p} n_i / s_i^2} \right)^2}.$$

The R function that performs the relevant computations, with arguments of interest, is

```
oneway.test(formula, data, subset, var.equal = FALSE)
```

The arguments `formula`, `subset` and `data` serve the same purpose as described for the `lm` function.[14]

Example

For purposes of comparison, apply Welch's test to the previous example.

```
> oneway.test(formula = score ~ age, data = algebraScores,
+       var.equal = FALSE)
```

to get $F^* = 74.361$, $\nu_1 = 4$, $\nu_2 = 168.18$ and p-value $= 0$. Observe that even though the number of factor levels is greater than 4, the results do not contradict those of the previous example.

15.5.3   Hotelling's T^2 Test

Hotelling's T^2 test, developed in [72], and as discussed in [92] and [93], is applicable to a balanced design in which samples from p populations, each containing at least p observations, are obtained. For $j = 1, 2, \ldots, p$, denote these p samples by $\{y_{1j}, y_{2j}, \ldots, y_{Ij}\}$ where $I \geq p$. The approach to performing the relevant computations for this procedure are an adaptation of ideas in [92] to the discussion presented in [74, Sec. 5.2].

Let $\mu_1, \mu_2, \ldots, \mu_p$ be the unknown true population means, then the procedure makes use of the fact that

$$\mu_1 = \mu_2 = \cdots = \mu_p \qquad \text{and} \qquad \mu_1 - \mu_2 = \mu_1 - \mu_3 = \cdots = \mu_1 - \mu_p = 0$$

are equivalent statements. With this in mind, for $k = 1, 2, \ldots, p - 1$, the $p - 1$ samples of pairwise differences $\{d_{1k}, d_{2k}, \ldots, d_{Ik}\}$ are obtained where, for $i = 1, 2, \ldots, I$, $d_{ik} = y_{i1} - y_{i(k+1)}$. Then, using $\bar{d}_k = \bar{y}_1 - \bar{y}_{k+1}$, compute

---

[14]Using the assignment `var.equal = TRUE` results in computations for the usual one-way ANOVA described previously.

the vector of pairwise mean differences $\bar{\mathbf{d}}' = (\bar{d}_1, \bar{d}_2, \ldots, \bar{d}_{p-1})$ and then, for $j, k = 1, 2, \ldots, p-1$, compute

$$s_{jk} = \frac{1}{I-1} \sum_{i=1}^{I} (d_{ij} - \bar{d}_j)(d_{ik} - \bar{d}_k),$$

the entries of the variance-covariance matrix, $\mathbf{S}$, associated with the samples of pairwise differences. The next step is to obtain the inverse of the matrix $\mathbf{S}$ and compute Hotelling's T^2 statistic,

$$T^2 = I\,\bar{\mathbf{d}}'\,\mathbf{S}^{-1}\,\bar{\mathbf{d}}.$$

Finally, since

$$F^* = \frac{(I-p+1)}{(p-1)(I-1)} T^2 \sim F(p-1, I-p+1),$$

see [74, p. 226], the F distribution with $\nu_1 = p-1$ and $\nu_2 = I-p+1$ degrees of freedom is used to calculate p-value $= P(F \geq F^*)$.

The computations for this procedure are performed by the function

```
hotellingT2.test(y, g, alpha = 0.05)
```

for which the arguments y and g are as for the `marascuiloMeans.test` function.

Example

Since a balanced design is required for this procedure, consider applying it to the dataset `PlantGrowth` from package `dataset`. These data, sourced from [41], contain results from an experiment conducted to compare dried weight yields of plants obtained under a control and two different treatments. Assume, for the sake of this example, that the constant variances assumption is not satisfied.

```
> with(data = PlantGrowth,
+      expr = hotellingT2.test(y = weight, g = group))
```

to get $F^* = 4.326$, $\nu_1 = 2$, $\nu_2 = 8$ and p-value $= 0.05328$. Since the constant variances assumption for these data is not quite marginal (see the script for this section), it is worth comparing the p-value from this procedure to those obtained by using Marascuilo's equality of means test, Welch's procedure and the traditional one-way F-test. It is found that Hotelling's procedure produces the largest p-value and Marascuilo's test produces the smallest, while the other two produce p-values that are close in magnitude.

15.6   Pairwise Comparisons: Unequal Variances

Three methods are provided here for cases where the equality of variances cannot be reasonably assumed. One of these is an extension of Marascuilo's equality of means test and one is a modification of Tukey's pairwise test of means that was proposed by Dunnett. The third, also proposed by Dunnett, involves the application of the studentized maximum modulus distribution.

15.6.1   Large-Sample Chi-Square Test

This approach is an extension of the procedure described in Section 15.5.1 to testing simple linear contrasts, see [87, Ex. 4]. For purposes of identification, this test will be referred to as *Marascuilo's pairwise comparison of means*.

With notation as already defined, $100(1 - \alpha)\%$ simultaneous confidence intervals are computed using

$$(\bar{y}_j - \bar{y}_k) - \sqrt{\chi_\alpha^2}\sqrt{\frac{s_j^2}{n_j} + \frac{s_k^2}{n_k}} < \mu_j - \mu_k < (\bar{y}_j - \bar{y}_k) + \sqrt{\chi_\alpha^2}\sqrt{\frac{s_j^2}{n_j} + \frac{s_k^2}{n_k}},$$

where χ_α^2 is such that $P\left(\chi^2 \geq \chi_\alpha^2\right) = \alpha$ with $p - 1$ degrees of freedom. Alternatively, using the test statistic

$$\chi_{jk}^{2*} = \frac{(\bar{y}_j - \bar{y}_k)^2}{s_j^2/n_j + s_k^2/n_k} \sim \chi^2(p - 1)$$

one may compute p-value $= P\left(\chi^2 \geq \chi_{jk}^{2*}\right)$ and compare the results against the joint significance level α.[15]

The function `marascuiloMPairs.test` performs the needed computations, and its usage definition is

```
marascuiloMPairs.test(y, g, alpha = 0.05, control = NULL)
```

The first three arguments for this function are as defined for the `marascuiloMeans.test` function; `y` is assigned the vector of observed responses, `g` the factor identifying the corresponding treatments and `alpha` the desired significance level. When the argument `control` is left in the default setting this function outputs results for all $m = p(p - 1)/2$ pairwise comparisons. If `control` is assigned a character string that identifies one of the levels of `g`, then this is treated as the control treatment and only $m = p - 1$ comparisons with the control are performed.

---

[15] A small sample adjustment is alluded to in [87], and using reasoning analogous to the global test, one may apply simultaneous Welch's two sample t tests along with a Bonferroni adjustment. See code for this section in the script file for an illustration.

Example

In the example from Section 15.5.1, the data suggest that statistically significant differences exist between the means of scores obtained by students from different age groups on their first exam in a GER mathematics course they are taking for the first time. To identify the statistically significant differences in mean scores run (output not shown)

```
> with(data = algebraScores,
+     expr = marascuiloMPairs.test(y = score, g = age))
```

From the results it is seen that the only pair of age groups that *does not* exhibit a statistically significant difference in mean scores is $[18, 22)$ versus $[22, 26)$.

15.6.2   Dunnett's C Procedure

This procedure was introduced in [44] as a modification of Tukey's method to accomodate unequal variances. Again, using previously defined notation, the formula for the $100(1 - \alpha)\%$ confidence interval for any pairwise difference $\mu_j - \mu_k$ is

$$(\bar{y}_j - \bar{y}_k) - q_{jk,\alpha} \sqrt{\frac{s_j^2}{n_j} + \frac{s_k^2}{n_k}} < \mu_j - \mu_k < (\bar{y}_j - \bar{y}_k) + q_{jk,\alpha} \sqrt{\frac{s_j^2}{n_j} + \frac{s_k^2}{n_j}},$$

with

$$q_{jk,\alpha} = \frac{q_{j,\alpha}\, s_j^2/n_j + q_{k,\alpha}\, s_k^2/n_k}{\sqrt{2}\left(s_j^2/n_j + s_k^2/n_k\right)}$$

where $q_{j,\alpha}$ and $q_{k,\alpha}$ are such that $P(q_j \geq q_{j,\alpha}) = \alpha$ with $q_j \sim q(p, n_j - 1)$ and $P(q_k \geq q_{k,\alpha}) = \alpha$ with $q_k \sim q(p, n_k - 1)$, respectively. Differences are considered statistically significant at the α level of significance if zero lies outside the interval.

Computations to obtain confidence intervals with this method are performed by the function

```
dunnettC.test(y, g, alpha = 0.05, control = NULL)
```

where all of the arguments are as for the `marascuiloMPairs.test` function.

Example

Revisiting the example from Section 15.5.1 to identify statistically significant differences using Dunnett's C procedure at $\alpha = 0.05$, and for all pairwise differences, run (output not shown)

```
> with(data = algebraScores,
+     expr = dunnettC.test(y = score, g = age))
```

Here too the only interval that has bounds of opposite signs, and hence the only pair of age groups that *does not* exhibit a statistically significant difference in mean scores is $[22, 26)$ versus $[18, 22)$.

15.6.3 Dunnett's T3 Procedure

This procedure was also introduced in [44]. The formula for the $100(1 - \alpha)\%$ confidence interval for any pairwise difference $\mu_j - \mu_k$ is

$$(\bar{y}_j - \bar{y}_k) - m_{jk,\alpha} \sqrt{\frac{s_j^2}{n_j} + \frac{s_k^2}{n_k}} < \mu_j - \mu_k < (\bar{y}_j - \bar{y}_k) + m_{jk,\alpha} \sqrt{\frac{s_j^2}{n_j} + \frac{s_k^2}{n_j}},$$

where $m_{jk,\alpha}$ is the studentized maximum modulus distribution quantile satisfying $P(M \geq m_{jk,\alpha}) = \alpha$ with $M \sim M(r, \nu_{jk})$ for which $r = p(p-1)/2$ and

$$\nu_{jk} = \frac{\left(s_j^2/n_j + s_k^2/n_k\right)^2}{\left(s_j^2/n_j\right)^2 \big/ (n_j - 1) + \left(s_k^2/n_k\right)^2 \big/ (n_k - 1)}.$$

Computations for this procedure are performed by the function

```
dunnettT3.test(y, g, alpha = 0.05, control = NULL)
```

where the arguments serve the same purposes as for the `marascuiloMPairs.test` and `dunnettC.test` functions.

Example

Once again, revisiting the example from Section 15.5.1, to determine the nature of the differences using Dunnett's T3 procedure at $\alpha = 0.05$ and for all pairwise differences, run (output not shown)

```
> with(data = algebraScores,
+        expr = dunnettT3.test(y = score, g = age))
```

As with the previous two illustrations, the only pair of age groups that *does not* exhibit a statistically significant difference in mean scores is $[22, 26)$ versus $[18, 22)$.

15.6.4 Comparisons with a Control

Any one of the three previously covered methods can be used for this purpose, with the understanding that the focus is on only the comparisons with the control treatment. The three functions, `marascuiloMPairs.test`, `dunnettC.test` and `dunnettT3.test`, can be instructed to focus on only these comparisons by identifying the control treatment through the argument `control` in the function call.

Example

Consider using the age group $[18, 22)$ as the control group, and suppose the multiple comparison procedure chosen to be used is Marascuilo's pairwise comparisons of means.

It is important to provide the function `marascuiloMPairs.test` the correct level label that identifies the control so, as a refresher, run

```
> levels(algebraScores$age)
[1] "[- ,18)" "[18,22)" "[22,26)" "[26,30)" "[30, +)"
```

Then the needed restricted computations can be performed by running (output not shown)

```
> with(data = algebraScores,
+     expr = marascuiloMPairs.test(y = score, g = age,
+         control = "[18,22)"))
```

The only difference in treatment means from the control that is not statistically significant is that between the $[18, 22)$ and $[22, 26)$ age groups.

The functions `dunnettC.test` and `dunnettT3.test` can be used similarly.

15.6.5 Which Procedure to Choose

All three interval formulas have the appearance

$$(\bar{y}_j - \bar{y}_k) - \omega_{jk}\sqrt{\frac{s_j^2}{n_j} + \frac{s_k^2}{n_k}} < \mu_j - \mu_k < (\bar{y}_j - \bar{y}_k) + \omega_{jk}\sqrt{\frac{s_j^2}{n_j} + \frac{s_k^2}{n_k}}$$

where the multiplier ω_{jk} depends on the method in question as defined in the previous sections. One may perform an informal comparison of the magnitudes of these multipliers in a manner similar to that done in Section 15.4.5 for the equal variances pairwise comparison methods. However, there are complications here. For a fixed significance level, the magnitude of each multiplier depends on one or more quantities. While $\sqrt{\chi_\alpha^2}$ remains fixed for a given number of means, p, the magnitudes of both $q_{jk,\alpha}$ and $m_{jk,\alpha}$ depend on the sample sizes *and* the sample variances. It is possible to prepare code for an informal graphical exploration of the behavior of the multipliers; see the script file.

More formally, in a simulation study of seven methods Moder [93] identified the Hotelling's T^2 test to be the best approach for balanced designs. For unbalanced designs it is suggested that an approach best suited for the given situation be chosen. A partial list of approaches can be found, for example, in [93] and [123, p. 550]. These observations concur with findings in [44] where analyses of four pairwise multiple comparison procedures, including the T3 and C procedures, are detailed.

15.7   The Nature of Differences Found

On performing any one of the procedures described in Sections 15.3 and 15.5, suppose the matter is to interpret the findings beyond simply identifying statistically significant differences. There are two scenarios: It may be that all possible pairwise comparisons are of interest. In his article [133], Tukey discussed this matter at length. For the case where a collection of treatments are compared with a control see, for example, Dunnett's discussion in [43, Sec. VII].

15.7.1   All Possible Pairwise Comparisons

Suppose the effects of p treatments are being analyzed through pairwise comparisons of the means $\mu_1, \mu_2, \ldots, \mu_p$, then for this scenario there will be $m = p(p-1)/2$ comparisons of pairs μ_j and μ_k where $j, k = 1, 2, \ldots, k$ and $j \neq k$. The intent of the analyses may be to simply obtain what are referred to as *homogeneous subsets*, or one may wish to go one step further and identify that homogeneous subset which corresponds to the *optimizing treatment(s)*.

In the current context, a homogeneous subset is a collection of means, all of whose pairwise differences are *not* found to be statistically significant. An optimizing treatment is one which has the effect of maximizing or minimizing the mean response, depending on the desired outcome. Here are three examples, the first two being reasonably straightforward and the third a bit more involved.

Example

In Section 15.5.1 pairwise comparisons of exam scores for students from different age-groups indicate statistically significant differences at $\alpha = 0.05$ for all pairs *except* for the $[18, 22)$ versus $[22, 26)$ pair. So, the homogeneous subsets are

$$\left\{ \mu_{[-,18)} \right\}, \ \left\{ \mu_{[18,22)}, \mu_{[22,26)} \right\}, \ \left\{ \mu_{[26,30)} \right\} \ \text{and} \ \left\{ \mu_{[30,+)} \right\}.$$

Next, by sorting the sample means in ascending order (see the script file for the code) the data suggest that, on average, students within the 18 to 26 year range score lowest on their first exam in a GER mathematics course they are taking for the first time, whereas students in the $[30, +)$ age range perform the best.

Example

In Section 15.4 pairwise comparisons of mean final exam scores for project-based students from the mathematical, natural, and social sciences gave

Pair	μ_{Math} vs. μ_{Sci}	μ_{Math} vs. μ_{Soc}	μ_{Sci} vs. μ_{Soc}
p-value	1.000	0.300	0.084

Thus, at the $\alpha = 0.10$ level of significance, μ_{Math} does not differ from μ_{Sci} and μ_{Soc}; however, μ_{Sci} does differ from μ_{Soc}. So, it seems reasonable to identify the homogeneous subsets as being

$$\{\mu_{Math}, \mu_{Sci}\} \text{ and } \{\mu_{Math}, \mu_{Soc}\}.$$

Observe that these sets are not mutually exclusive, both contain μ_{Math}, and this might give rise to some unease. One may alleviate this discomfort by observing that $\bar{x}_{Math} = 83.43$ is closer in value to $\bar{x}_{Sci} = 83.80$ than to $\bar{x}_{Soc} = 80.41$, and then consider, instead, the mutually exclusive subsets

$$\{\mu_{Math}, \mu_{Sci}\} \text{ and } \{\mu_{Soc}\}.$$

So, one might conclude that higher mean scores for the project-based course are associated with students from the mathematical and natural sciences. However, it is important not to ignore the fact that the difference between μ_{Math} and μ_{Soc} was not found to be statistically significant.

Example

The previous two examples are straightforward for one of two reasons. In the first example the number of non-significant differences in means is low, so gathering means into homogeneous subsets is quite manageable. For the second example the grouping of means into homogeneous subsets is fairly easy because of the small number of pairwise comparisons involved.

Here, consider a case where the number of pairwise comparisons is large *and* the number of non-significant differences is also large. The data frame `chickwts` contained in package `datasets` provides an example of such a case.

These data are from an experiment conducted to measure and compare the effectiveness of various feed supplements on the growth rate of chickens. More about this dataset can be found by running `?chickwts`, and a quick summary of the data can be obtained by running `summary(chickwts)`.

Continuing, fit the data to an `aov` object then perform all pairwise comparisons using Tukey's HSD procedure at the default 95% confidence level (output not shown),

```
> mod <- aov(formula = weight ~ feed, data = chickwts)
> round(TukeyHSD(x = mod, ordered = TRUE)$feed, 4)
```

Finally, sort the sample means in ascending order to get a feel for the relative sizes of sample means and to aid in organizing homogeneous subsets, possibly representing them graphically as shown in Figure 15.5.

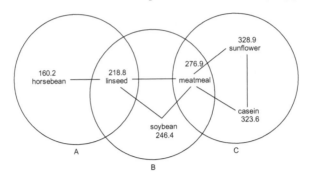

FIGURE 15.5
Graphical representation of non-significant differences between mean weights.
Sample means are included above each feed type and line segments between
feed types identify non-significant differences. The circles A, B, and C enclose
homogeneous subsets of feed types; note that as displayed here some of the
subsets are not mutually exclusive.

Then, focusing on the feed types linseed and meatmeal, and paying attention to the relative magnitudes of differences in their means from neighboring
feed types, one might identify the mutually exclusive homogeneous subsets as
being

{horsebean}, {linseed, soybean, meatmeal}, and {casein, sunflower}.

Thus, one might conclude that the data suggest that the feed types that
maximize weight gain are casein and sunflower, with the acknowledgement
that the differences between casein and meatmeal, and between linseed and
horsebean are not statistically significant.

15.7.2 Comparisons with a Control

Suppose the kth treatment is the control treatment, then this scenario involves
$m = p - 1$ comparisons of the means μ_j with μ_k and $j \neq k$. A goal might be
to identify those treatment(s) whose effects result in the largest statistically
significant difference(s) from the control *in the desired direction* (left/negative
estimates versus right/positive estimates).

For example, in the Section 15.6.4 illustration the only pair of age groups
that *does not* exhibit a statistically significant difference in mean scores from
the control group, $[18, 22)$, is $[22, 26)$. The fact that all statistically significant
differences (from the control) are positive suggests that the mean scores for the
remaining age groups are on average greater than that of the control group,
the largest difference being for the $[30, +)$ age group.

16

Selected Tests for Medians and More

This chapter covers a selection of methods applicable to data for which distributional assumptions on the underlying random variables for earlier encountered methods are either not satisfied, or for which the distributions are unknown. Methods include those applicable to data in which the response variable is numeric and the median is of interest, as well as those applicable to data in which the response variable is ordinal. Code for all illustrations, and more, are contained in the script file, `Chapter16Script.R`.

16.1 Relevant Probability Distributions

The three previously introduced distributions that find use in some of the tests covered in this chapter include the binomial, standard normal and the chi-square distributions. Two additional discrete distributions that are applicable to exact tests presented in this chapter are those of the Signed Rank statistic and of the Rank Sum statistic.

16.1.1 Distribution of the Signed Rank Statistic

Let $y_1, y_2, \ldots, y_n$ be a random sample of nonzero observations from a symmetric continuous distribution. For $i = 1, 2, \ldots, n$, let r_i denote the rank of $|y_i|$ in the set $|y_1|, |y_2|, \ldots, |y_n|$, with tied cases being assigned average ranks. Then the *Wilcoxon signed rank statistic* T^+ is the sum of the ranks r_i for which $y_i > 0$. In the absence of ties among the absolute values, this statistic takes on integer values from 0 to $n(n+1)/2$ and it is worth noting that, for any quantile T^* in this range,

$$P(T^+ \leq T^*) = P(T^+ \geq n(n+1)/2 - T^*).$$

The mean and variance of the underlying random variable are

$$\mu_{T^+} = \frac{n(n+1)}{4} \quad \text{and} \quad \sigma^2_{T^+} = \frac{n(n+1)(2n+1)}{24},$$

respectively, and quantities possessing this distribution will be identified using notation of the form $T^+(n)$.

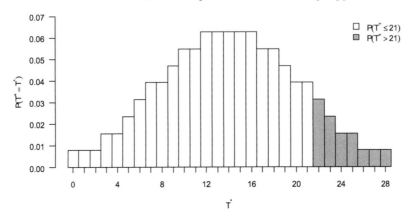

FIGURE 16.1

Density plot of $T^+(7)$ using the function `dsignrank`. The horizontal axis represents the sum of the ranks corresponding to positive observations in a random sample, $y_1, y_2, \ldots, y_7$, of nonzero observations. See the script file for code used to construct this figure.

The two R functions of interest for this distribution are `psignrank` and `qsignrank`, contained in package `stats`. Use

```
psignrank(q, n, lower.tail = TRUE)
```

for computing a left-tailed probability $P(T^+ \leq T^*)$ for a given quantile T^*. Setting `lower.tail = FALSE` permits calculations associated with right-tailed probabilities of the form $P(T^+ > T^*)$.

If, in the presence of ties, T^* is not an integer, then the function `psignrank` approximates probabilities by rounding T^* *up* to the nearest integer. That is,

$$P(T^+ \leq T^*) \approx P(T^+ \leq \lceil T^* \rceil) \quad \text{and} \quad P(T^+ > T^*) \approx P(T^+ > \lceil T^* \rceil).$$

As for previously seen quantile estimating functions for discrete distributions, the smallest quantile T^* which satisfies $P(T^+ \leq T^*) \geq P$ for a given probability P is obtained using

```
qsignrank(p, n, lower.tail = TRUE)
```

Setting `lower.tail = FALSE` obtains the smallest quantile T^* for which $P(T^+ > T^*) < P$. Examples similar to those given in Section 11.1.1 to illustrate these are given in the script file for this section.

As for previously described distribution functions, the argument `q` is assigned a quantile or a vector of quantiles and `p` is assigned a probability or a vector of probabilities. For applications in this chapter the argument n is assigned the number of nonzero observations in the sample under investigation.

Examples

Consider the distribution $T^+(7)$, the density plot of which is shown in Figure 16.1. To calculate $P(T^+ \leq 21) = 0.890625$, run

```
> psignrank(q = 21, n = 7, lower.tail = TRUE)
```

and to compute $P(T^+ > 21) = 0.109375$, run

```
> psignrank(q = 21, n = 7, lower.tail = FALSE)
```

Keep in mind that $P(T^+ > 21) = P(T^+ \geq 22)$ since $T^+(n)$ is a discrete distribution.

To find the smallest quantile $(T^* = 4)$ that satisfies $P(T^+ \leq T^*) \geq 0.05$, run

```
> qsignrank(p = 0.05, n = 7, lower.tail = TRUE)
```

To find the smallest quantile $(T^* = 24)$ that satisfies $P(T^+ > T^*) > 0.05$, run

```
> qsignrank(p = 0.05, n = 7, lower.tail = FALSE)
```

Remembering that $P(T^+ \geq T^*) = P(T^+ > T^* - 1)$ and the symmetry property

$$P(T^+ \leq T^*) = P(T^+ \geq n(n+1)/2 - T^*),$$

observe that $P(T^+ \leq 4) = P(T^+ \geq 24)$.

16.1.2 Distribution of the Rank Sum Statistic

There are two equivalent descriptions associated with this statistic. Consider, first, the *Mann–Whitney U-statistic* since R functions for such distributions are available in package `stats`.

The Mann–Whitney U-Statistic

Let $x_1, x_2, \ldots, x_m$ and $y_1, y_2, \ldots, y_n$ be two random and independent samples having continuous underlying random variables.[1] Then if all elements of the two samples are distinct, for $i = 1, 2, \ldots, m$ and $j = 1, 2, \ldots, n$, the test statistic U is the number of pairs (x_i, y_j) for which $x_i > y_j$. In this case, and

---

[1]Some arrange the first and second samples such that $m \leq n$ (see, for example, [77, Sec. 14.3]), and others such that $n \leq m$ (see, for example, [70, p. 108]). This is a table dependent requirement and is not needed if the available R probability distribution functions are used instead of tables.

for all possible scenarios, this statistic takes on integer values from 0 to mn, and in a manner similar to the previous Sign Rank distribution,

$$P(U \leq U^*) = P(U \geq mn - U^*).$$

The mean and variance of U are

$$\mu_U = \frac{mn}{2}, \quad \text{and} \quad \sigma_U^2 = \frac{mn(m+n+1)}{12},$$

respectively, and quantities with this distribution will be identified using notation of the form $U(m, n)$.

Left-tailed probabilities of the form $P(U \leq U^*)$ are computed using

```
pwilcox(q, m, n, lower.tail = TRUE)
```

and, as is usual, setting `lower.tail = FALSE` permits calculations associated with right-tailed probabilities of the form $P(U > U^*)$.

In the presence of ties, the U statistic is generalized in the following manner. For $i = 1, 2, \ldots, m$ and $j = 1, 2, \ldots, n$ first obtain

$$\delta_{ij} = \begin{cases} 1 & \text{if } x_i > y_j \\ 1/2 & \text{if } x_i = y_j \\ 0 & \text{if } x_i < y_j \end{cases}$$

Then, for the given samples,

$$U^* = \sum_{i=1}^{m} \sum_{j=1}^{n} \delta_{ij},$$

and, if U^* is not an integer, `pwilcox` approximates probabilities by rounding U^* *down* to the nearest integer. That is,

$$P(U \leq U^*) \approx P(U \leq \lfloor U^* \rfloor) \quad \text{and} \quad P(U \geq U^*) \approx P(U \geq \lfloor U^* \rfloor).$$

Moving to the quantile function, given a left-tailed probability P, to estimate the smallest quantile U^* for which $P(U \leq U^*) \geq P$, use

```
qwilcox(p, m, n, lower.tail = TRUE)
```

If `lower.tail = FALSE` is used, then the smallest quantile U^* for which $P(U > U^*) < P$ is found. The arguments q and p are used as for the previous Sign Rank functions, and the arguments m and n are assigned the number of observations in the first and second samples, respectively.

The Wilcoxon W-Statistic

This statistic is obtained by combining the two samples and then arranging and ranking the entries of the combined sample in ascending order; tied cases

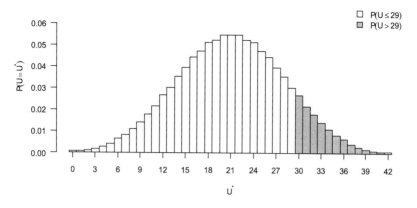

FIGURE 16.2
Density plot of $U(6,7)$ obtained using the function `dwilcox`. For random samples $x_1, x_2, \ldots, x_6$ and $y_1, y_2, \ldots, y_7$ with distinct elements, the horizontal axis represents the number of pairs (x_i, y_j) for which $x_i > y_j$.

are assigned average ranks. The *Wilcoxon W-statistic* is then obtained by summing the resulting ranks associated with values from the first sample.[2] It turns out (see, for example, [70, pp 117–118] or [7, p. 484]) that

$$W = U + \frac{m(m+1)}{2};$$

that, in the absence of ties, W takes on integer values from $m(m+1)/2$ to $m(m+2n+1)/2$; and in this case

$$P(W \leq W^*) = P(W \geq m(m+n+1) - W^*).$$

The mean and variance of W are

$$\mu_W = \frac{m(m+n+1)}{2}, \quad \text{and} \quad \sigma_W^2 = \frac{mn(m+n+1)}{12},$$

respectively, and quantities with this distribution will be identified using notation of the form $W(m, n)$.

It is a simple matter of translating W to U to take advantage of the functions `pwilcox` and `qwilcox` to perform computations associated with probabilities of the Wilcoxon W-statistic.

---

[2]Some, for example, [70, p. 107] obtain the statistic by summing the ranks associated with the second sample. The definition of the W-statistic used here is chosen to conform with the functions `pwilcox` and `qwilcox`, and the definition used in the later demonstrated `wilcox.test` function.

Examples Involving the U-Statistic

Consider the distribution $U(6,7)$, the density plot of which is given in Figure 16.2. To compute $P(U \leq 29) = 0.8828671$, run

```
> pwilcox(q = 29, m = 6, n = 7, lower.tail = TRUE)
```

and, to compute $P(U > 29) = 0.1171329$, run

```
> pwilcox(q = 29, m = 6, n = 7, lower.tail = FALSE)
```

Next, to find the smallest quantile $(U^* = 9)$ for which $P(U \leq U^*) \geq 0.05$ use

```
> qwilcox(p = 0.05, m = 6, n = 7, lower.tail = TRUE)
```

and, to find the smallest quantile $(U^* = 33)$ for which $P(U > U^*) < 0.05$ use

```
> qwilcox(p = 0.05, m = 6, n = 7, lower.tail = FALSE)
```

See the script file for a more detailed examination of the output obtained from these two `qwilcox` function calls.

Examples Involving the W-Statistic

Consider the distribution $W(6,7)$, the density plot of which is a horizontal translation of the plot given in Figure 16.2 $m(m+1)/2 = 21$ units to the right. Then, since
$$W^* = U^* + m(m+1)/2,$$
a transformation of $W^* = 50$ shows that

$$P(W \leq 50) = P(U \leq 29) \quad \text{and} \quad P(W > 50) = P(U > 29),$$

and the code in the previous example does the job. From the previous example,

```
> qwilcox(p = 0.05, m = 6, n = 7, lower.tail = TRUE)
```

gives the smallest quantile $(U^* = 9)$ for which $P(U \leq U^*) \geq 0.05$ and, consequently, with $m = 6$ gives $W^* = U^* + m(m+1)/2 = 30$, the smallest W^* for which $P(W \leq W^*) \geq 0.05$. The smallest W^* for which $P(W > W^*) < 0.05$ can be similarly found.

16.2   The One-Sample Sign Test

This test, an extension of the one-proportion test from Section 11.2, can be used on single-sample data to determine if a population median differs significantly from a hypothesized median.

Let $y_1, y_2, \ldots, y_N$ be a random sample for which the underlying random variable Y is either not normally distributed, or whose distribution is unknown. Let $\tilde{\mu}$ denote the true (unknown) population median and $\tilde{\mu}_0$ the hypothesized median, then the possible alternative hypotheses include:

$$H_1 : \tilde{\mu} < \tilde{\mu}_0; \qquad H_1 : \tilde{\mu} \neq \tilde{\mu}_0; \qquad \text{or} \qquad H_1 : \tilde{\mu} > \tilde{\mu}_0.$$

For this method, observations that equal $\tilde{\mu}_0$ are excluded from the test and the number of observations, m^*, in the data that are greater in value than $\tilde{\mu}_0$ is found.[3]

16.2.1 The Exact Test

Under the null hypothesis $m^* \sim BIN(n, p)$ where n represents the number of observations that do not equal the hypothesized median. The probability that a randomly observed Y is greater than the hypothesized median is taken to be $p = 0.5$ since the median marks the 50th percentile of any distribution. So, the p-value is obtained using $Y \sim BIN(n, p)$ according to the following conventions (see, for example, [27, pp. 158–160]):

$$\text{p-value} = \begin{cases} P(Y \leq m^*) & \text{for a left-tailed test; and} \\ P(Y \geq m^*) & \text{for a right-tailed test;} \end{cases}$$

and for a two-tailed test,

$$\text{p-value} = \begin{cases} 2P(Y \leq m^*) & \text{if } P(Y \leq m^*) \leq 0.5, \text{ and} \\ 2P(Y \geq m^*) & \text{if if } P(Y \leq m^*) > 0.5. \end{cases}$$

Two approaches to perform the calculations can be followed.

Example

For a generic sample, `signTestOne`, consider testing the hypotheses

$$H_0 : \tilde{\mu} \geq 27 \quad \text{vs.} \quad H_1 : \tilde{\mu} < 27.$$

Then, basic code along with the **pbinom** function can be used to perform the necessary computations, for example,

```
> n <- length(signTestOne[signTestOne != 27])
> mStar <- length(signTestOne[signTestOne > 27])
> pbinom(q = mStar, size = n, prob = 0.5)
```

---

[3]That is, m^* is the number of cases for which $y_i - \tilde{\mu}_0$, $i = 1, 2, \ldots, N$ has a positive sign.

gives p-value $= 0.03195733$.

The function `binom.test` (see Section 11.2.5) duplicates this result. For here the usage definition, with arguments of interest, is

```
binom.test(x, n, p = 0.5,
         alternative = c("two.sided", "less", "greater"))
```

The argument `n` is assigned the number of observations that do not equal the hypothesized median, the hypothesized proportion `p` is left with the default value, and for purposes of the Sign Test, `x` is assigned m^*, the number of observations in the sample that are greater than the hypothesized median.

Either way, at any $\alpha \geq 0.032$, there is evidence to suggest that the true median is less than 27.

16.2.2   The Normal Approximation

As for the one-proportion test in Section 11.2, a normal approximation to the above exact test can be used for large samples. For example, in [27, Sec. 3.4] it is suggested that the normal approximation be used if $n > 20$. In this case the test statistic becomes equivalent to the one described in Section 11.2.3, with $p_0 = 0.5$, which through some simplification and including a correction for continuity results in[4]

$$z^* = \frac{2m^* - n + \delta}{\sqrt{n}},$$

where

$$\delta = \begin{cases} 1 & \text{for a left-tailed probability, and} \\ -1 & \text{for a right-tailed probability.} \end{cases}$$

In this case the p-value is obtained using the standard normal distribution according to the conventions:

$$\text{p-value} = \begin{cases} P(Z \leq z^*) & \text{for a left-tailed test; and} \\ P(Z \geq z^*) & \text{for a right-tailed test;} \end{cases}$$

and, for a two-tailed test,

$$\text{p-value} = \begin{cases} 2P(Z \leq z^*) & \text{if } P(Z \leq z^*) \leq 0.5; \text{ and} \\ 2P(Z \geq z^*) & \text{if } P(Z \leq z^*) > 0.5. \end{cases}$$

Here too, the necessary computations require just a few lines of basic code.

Example

Since $n > 20$ for the data in `signTestOne`, the normal approximation can be applied in testing the same hypotheses. Including a correction for continuity, running

---

[4]If a correction for continuity is not used $z^* = (2m^* - n)/\sqrt{n}$.

```
> zStar <- (2*mStar - n + 1)/sqrt(n)

> pnorm(q = zStar, lower.tail = TRUE)
```

produces a p-value $= 0.03309629$, which is close in value to that obtained using the exact test.

The function `prop.test` (see Section 11.2.4) can also be used to obtain an identical p-value by performing the associated chi-square test.[5] The usage definition for this function as applied here is

```
prop.test(x, n, p = 0.5, correct = TRUE,

    alternative = c("two.sided", "less", "greater"))
```

and all arguments serve the same purpose as for the previously described `binom.test` function. See the script file for illustrations.

16.3   Paired Samples Sign Test

The single sample Sign Test can be used on paired (dependent) samples to determine if there is a difference between the population medians.

For $j = 1, 2$ let $y_{1j}, y_{2j}, \ldots, y_{N_j}$ be random dependent samples for which the underlying random variables Y_j are either not normally distributed, or whose distributions are unknown. Let $\tilde{\mu}_1$ and $\tilde{\mu}_2$ denote the true (unknown) medians, then the possible alternative hypotheses include:

$$H_1 : \tilde{\mu}_1 < \tilde{\mu}_2; \qquad H_1 : \tilde{\mu}_1 \neq \tilde{\mu}_2; \qquad \text{or} \qquad H_1 : \tilde{\mu}_1 > \tilde{\mu}_2.$$

Since the samples are paired, it suffices to consider the sample of paired differences, $d_i = y_{i1} - y_{i2}$, for $i = 1, 2, \ldots, N$, and the equivalent alternative hypotheses:

$$H_1 : \tilde{\mu}_d < 0; \qquad H_1 : \tilde{\mu}_d \neq 0; \qquad \text{or} \qquad H_1 : \tilde{\mu}_d > 0.$$

The task then becomes equivalent to a single sample Sign test performed on

---

[5]The test statistic formula used for one-tailed tests in the function `prop.test`, including a correction for continuity, is equivalent to

$$z^* = \delta \frac{|2m^* - n| - 1}{\sqrt{n}}$$

where

$$\delta = \begin{cases} 1 & \text{if } 2m^* - n \geq 0, \\ -1 & \text{if } 2m^* - n < 0. \end{cases}$$

It is useful to be aware that under certain rare cases this results in p-values that differ from those obtained using the above described normal approximation approach. If a correction for continuity is not included the two approaches produce identical results.

the sample of paired differences, $d_1, d_2, \ldots, d_N$, with a hypothesized median of zero. Consequently, paired differences that equal zero are excluded from the test and m^*, the number of paired differences among the remaining $n \leq N$ differences, that are positive is found. The exact test and normal approximation approaches are as described for single sample Sign Test in the previous section.

Example: Exact Test

Consider testing the hypotheses

$$H_0 : \tilde{\mu}_1 = \tilde{\mu}_2 \quad \text{vs.} \quad H_1 : \tilde{\mu}_1 \neq \tilde{\mu}_2$$

for the paired samples (y1 and y2) contained in the data frame signTestTwo. Once the paired differences, d_i, are found by running, for example,

```
> d <- with(data = signTestTwo, expr = y1 - y2)
```

n and m^* and p-value $= 0.2477886$ are found by running code similar to the one-sample case. So, the difference in medians is not statistically significant, at least for typical levels of significances. This result is duplicated by the binom.test function.

Example: The Normal Approximation

This is simply a matter of using n and m^* to find corrected z^*-values for left- and right-tailed probabilities, and then using the standard normal distribution to compute p-value $= 0.2482131$. In this case the result is duplicated by the prop.test function (see the script file).

16.4   Independent Samples Median Test

For $j = 1, 2, \ldots, J$ where $J \geq 2$, let $y_{1j}, y_{2j}, \ldots, y_{N_j}$ be random independent samples for which the underlying random variables Y_j are either not normally distributed, or whose distributions are unknown. Denote the medians of the J populations by $\tilde{\mu}_j$, and for purposes of this section, the interval measurement scale is used.

16.4.1   Equality of Medians

The task at hand is to test the hypotheses

$$H_0 : \tilde{\mu}_1 = \tilde{\mu}_2 = \cdots = \tilde{\mu}_J \quad \text{vs.} \quad H_1 : \text{The medians are not all equal.}$$

As described in [27, Sec. 4.3] Pearson's proportions test from Section 12.1.1 can be used to test these hypotheses in the following manner.

Under the null hypothesis, the J populations have the same median. Thus, under the null hypothesis all populations have the same probability of an observation exceeding the median of the combined populations. Denote the sample grand median by $\tilde{y}$, and obtain counts for the 2-dimensional contingency table.

	1	2	$\cdots$	J	Totals
Success $(y_{ij} > \tilde{y})$	o_{11}	o_{12}	$\cdots$	o_{1J}	r_1
Failure $(y_{ij} \leq \tilde{y})$	o_{21}	o_{22}	$\cdots$	o_{2J}	r_2
Totals	n_1	n_2	$\cdots$	n_J	n

The first row represents the frequencies of observed data from each sample that exceed the grand median, and the second row the frequencies of observed data that do not. This layout is identical to that of Section 12.1.1, and as is the case there the expected frequencies are computed using $e_{ij} = r_i\, n_j/n$. Then the test statistic is

$$\chi^{2*} = \sum_{i=1}^{2}\sum_{j=1}^{J} (o_{ij} - e_{ij})^2 / e_{ij}$$

and, if $e_{ij} \geq 5$ is true for every expected frequency, p-value $= P(\chi^2 \geq \chi^{2*})$ is obtained using $\chi^2(J - 1)$. The chisq.test function used in Section 12.1.1 can be used as illustrated in the following example.

Example

Recall the data frame elemStatStudy introduced and described in Section 12.1.1. Let $\tilde{\mu}_{\text{Mat}}$, $\tilde{\mu}_{\text{Nat}}$ and $\tilde{\mu}_{\text{Soc}}$ denote the true median scores of students from the mathematical sciences, natural sciences and social sciences, respectively, in traditionally taught courses. Consider testing the claim that the median scores for the three groups of students are the same using, say $\alpha = 0.10$.

The first step is to extract the relevant data and prepare it in the form of a 2×3 contingency table named, for example, obsFreq for use by the chisq.test function (see the script file). Then, running

```
> chisq.test(x = obsFreq)
```

yields $\chi^{2*} = 5.615$, $\nu = 2$, and p-value $= 0.06035$. So, at the chosen level of significance, there is evidence to reject the claim.[6]

---

[6]The prop.test function can also be used; see the script for this section.

16.4.2 Pairwise Comparisons of Medians

If, based on findings from the median test, there is evidence that statistically significant differences in medians exist, one might wish to determine the nature of these differences through simultaneous pairwise comparisons. In [27, p. 220] it is stated that this can be accomplished by performing pairwise median tests at the same level of significance as the original test is conducted. The function prop.test from Section 11.3.4, without a correction for continuity, works well for this purpose.

Example

Consider applying this strategy to the data in the contingency table obsFreq from the previous example. After extracting the needed frequencies and performing other relevant initializations, nested for-loops can be used to perform the relevant prop.test function calls. Finally, relevant test results can then be outputted in data frame format (see script file for complete code).

```
           Pair ChiSq df p.value
1 mathsci-natsci 0.036  1  0.8486
2 mathsci-socsci 2.807  1  0.0938
3  natsci-socsci 5.027  1  0.0250
```

The results indicate that, at the $\alpha = 0.10$ level of significance, the median score of the social sciences students differs from both of the other two groups.

16.5 Single Sample Signed Rank Test

This test differs from the Sign Test in that ranks are incorporated into the test. As is the case for the Sign Test, this test can be used on single-sample data to determine if a population median differs significantly from a hypothesized median.

Let $\tilde{\mu}$ denote the true (unknown) population median and $\tilde{\mu}_0$ the hypothesized median, then the possible alternative hypotheses to be tested include:

$$H_1 : \tilde{\mu} < \tilde{\mu}_0; \qquad H_1 : \tilde{\mu} \neq \tilde{\mu}_0; \qquad \text{or} \qquad H_1 : \tilde{\mu} > \tilde{\mu}_0.$$

If the underlying distribution is symmetric and a random sample of independent observations, $y_1, y_2, \ldots, y_N$, is obtained, then the test statistic for the above hypotheses is obtained as follows.

First, compute the sequence of differences $d_i = y_i - \tilde{\mu}_0$, $i = 1, 2, \ldots, N$, and exclude those cases for which the difference is zero. For the remaining $n \leq N$ cases, obtain the ranks r_i of the *absolute differences*, assigning mean ranks to tied cases. Finally, the test statistic, T^*, is obtained by the summing the ranks corresponding to cases for which $d_i > 0$. The idea being that if T^* is large (small) more observations lie above (below) the hypothesized median.

The R function for this test along with the relevant arguments is

```
wilcox.test(x, mu = 0, exact = NULL, correct = TRUE,
      alternative = c("two.sided", "less", "greater"))
```

The argument x is assigned the vector of observations that do not equal the hypothesized median and mu is assigned the hypothesized median $\tilde{\mu}_0$. The arguments exact and correct are left in the default setting (their purposes are explained in the examples that follow), and assignments to the argument alternative serve the usual purpose.

16.5.1 The Exact Test

If there are no ties in the n nonzero differences and $n < 50$, by default the function wilcox.test computes the exact p-value corresponding to T^* using the $T^+(n)$ distribution described in Section 16.1.1. Alternatively, the function pwilcox can be used to compute

$$\text{p-value} = \begin{cases} P(T^+ \leq T^*) & \text{for a left-tailed test; and} \\ P(T^+ \geq T^*) & \text{for a right-tailed test;} \end{cases}$$

and, for a two-tailed test

$$\text{p-value} = \begin{cases} 2P(T^+ \leq T^*) & \text{if } P(T^+ \leq T^*) \leq 0.5; \text{ and} \\ 2P(T^+ \geq T^*) & \text{if } P(T^+ \leq T^*) > 0.5. \end{cases}$$

Example

Consider testing the hypotheses

$$H_0 : \tilde{\mu} = 27 \quad \text{vs.} \quad H_1 : \tilde{\mu} \neq 27$$

on the data contained in signRankOne. A quick check shows that the data contain no ties, so the arguments exact and correct need not be included in the wilcox.test function call,

```
> wilcox.test(x = signRankOne[signRankOne != 27],
+     alternative = "two.sided", mu = 27)
```

to get p-value $= 0.5803$. So, at traditional levels of significance there is insufficient evidence to conclude that the true median differs from 27.

Note that, with respect to performing the exact test, the wilcox.test function does not like observations that equal the hypothesized median, hence the argument assignment x = signRankOne[signRankOne != 27].

16.5.2   The Normal Approximation

In [77, p. 691] it is mentioned that if the sample size satisfies $n \geq 20$, then in the absence of ties in the data T^+ is approximately normally distributed with mean and variance

$$\mu_{T^+} = \frac{n(n+1)}{4}, \quad \text{and} \quad \sigma^2_{T^+} = \frac{n(n+1)(2n+1)}{24},$$

respectively.

In [27, p. 353] it is suggested that the normal approximation be used if there are many ties or if $n > 50$. In the presence of ties, it is recommended in [70, pp. 38–39] that the variance of T^+ be adjusted as follows.

Let $g < n$ denote the number of distinct differences (Note: In the absence of tied differences $g = n$) and call these g distinct differences the tied groups. Next, for $j = 1, 2, \ldots, g$ let t_j be the number of tied differences in the jth tied group where, if the jth tied group has a single case, set $t_j = 1$. Then, compute

$$\sigma^2_{T^+} = \left[n(n+1)(2n+1) - \frac{1}{2} \sum_{j=1}^{g} t_j(t_j - 1)(t_j + 1) \right] \bigg/ 24.$$

Either way, the test statistic, including a correction for continuity, is

$$z^* = \frac{T^* - \mu_{T^+} + 0.5\,\delta}{\sigma_{T^+}},$$

where

$$\delta = \left\{ \begin{array}{ll} 1 & \text{for a left-tailed probability} \\ -1 & \text{for a right-tailed probability} \end{array} \right.$$

and the p-value is obtained as described in Section 16.2.2.

If there are any ties in the n nonzero differences the function `wilcox.test` computes the approximate p-value through the normal approximation just described.

Example: No Tied Differences Present

Code to implement the above described computations for the hypotheses

$$H_0 : \tilde{\mu} = 27 \quad \text{vs.} \quad H_1 : \tilde{\mu} \neq 27$$

on the sample y from the previous example is given in the script for this section. To apply the normal approximation to the same problem using the `wilcox.test` function, run

```
> wilcox.test(x = signRankOne[signRankOne != 27],
+      alternative = "two.sided",
+      correct = TRUE, exact = FALSE, mu = 27)
```

to get p-value = 0.5737. While the p-value obtained is slightly smaller than that obtained using the exact test, the result of the test remains unchanged.

Example: Tied Differences Present

The vector `signRankOneTies` is an expansion of `signRankOne` through an inclusion of some tied data. Consider testing the hypotheses

$$H_0 : \tilde{\mu} \geq 27 \quad \text{vs.} \quad H_1 : \tilde{\mu} < 27$$

using these data. Then,

```
> wilcox.test(x = signRankOneTies[signRankOneTies != 27],
+      alternative = "less",
+      correct = TRUE, exact = FALSE, mu = 27)
```

to get p-value = 0.1066. So, the alternative is supported for significance levels $\alpha \geq 0.1066$.

Note that even if the argument assignment `exact = FALSE` is not included in the function call, the normal approximation is still implemented, but with a warning to the user.

16.6 Paired Samples Signed Rank Test

As with the Sign Test, the Signed Rank Test can be applied to a sample of paired differences, $d_i = y_{i1} - y_{i2}$, for $i = 1, 2, \ldots, N$, and the relevant alternative hypotheses are equivalent to:

$$H_1 : \tilde{\mu}_d < 0; \qquad H_1 : \tilde{\mu}_d \neq 0; \qquad \text{or} \qquad H_1 : \tilde{\mu}_d > 0.$$

Here too the only difference between this and the one-sample test is that the d_i are paired differences and the hypothesized value is always zero. Absolute differences, $|d_i|$ (excluding zeros) are ranked, ties being assigned average ranks, and the test statistic, T^+, is obtained by the summing the ranks corresponding to positive differences (that is $d_i > 0$). The p-value is computed as previously described.

Example: No Tied Differences Present

Consider testing the hypotheses

$$H_0 : \tilde{\mu}_1 \leq \tilde{\mu}_2 \quad \text{vs.} \quad H_1 : \tilde{\mu}_1 > \tilde{\mu}_2$$

using the paired samples (`y1` and `y2`) contained in the data frame `signRankTwo`. A quick check show there are no ties in the absolute differences, as well as no zero-differences. Then, the exact test can then be performed by running

```
> with(data = signRankTwo, expr = wilcox.test(x = y1 - y2,
    mu = 0, alternative = "greater")
```

to get p-value = 0.7951. So, there is insufficient evidence at any typical significance level to support a conclusion that $\tilde{\mu}_1 > \tilde{\mu}_2$.[7]

Example: Tied Differences Present

Consider applying this method to the paired samples (y1 and y2) in the data frame `signRankTwoTies` to test hypotheses

$$H_0 : \tilde{\mu}_1 = \tilde{\mu}_2 \quad \text{vs.} \quad H_1 : \tilde{\mu}_1 \neq \tilde{\mu}_2.$$

A quick check shows that both tied-differences and zero-differences exist. Then, running

```
> with(data = signRankTwoTies,
+     expr = wilcox.test(x = y1, y = y2,
+         alternative = "two.sided",
+         paired = TRUE, exact = FALSE))
```

gives p-value = 0.4858.

16.7   Rank Sum Test of Medians

This test is commonly referred to as the *Mann–Whitney Test*, typically if the Mann–Whitney U statistic is used, or as the *Wilcoxon Rank Sum Test*, typically if the Wilcoxon W statistic is used. As the discussions on the Rank Sum statistic(s) in Section 16.1.2 suggest, the two approaches are equivalent.

The task is to compare medians associated with two independent population using two independent random samples, call these $y_{11}, y_{21}, \ldots, y_{m1}$ and $y_{12}, y_{22}, \ldots, y_{n2}$. Let $\tilde{\mu}_1$ and $\tilde{\mu}_2$ denote the true (unknown) population medians, respectively, then the possible alternative hypotheses include:

$$H_1 : \tilde{\mu}_1 < \tilde{\mu}_2; \quad H_1 : \tilde{\mu}_1 \neq \tilde{\mu}_2; \quad \text{or} \quad H_1 : \tilde{\mu}_1 > \tilde{\mu}_2.$$

The computational steps for each of the approaches are summarized below.

---

[7] Note that an equivalent function call would be

```
with(data = signRankTwo, expr = wilcox.test(x = y1, y =y2,
    alternative = "greater", paired = TRUE, correct = FALSE))
```

16.7.1   The Exact Mann–Whitney Test

First, for $i = 1, 2, \ldots, m$ and $j = 1, 2, \ldots, n$, define

$$\delta_{ij} = \begin{cases} 1 & \text{if } y_{i1} > y_{j2} \\ 1/2 & \text{if } y_{i1} = y_{j2} \\ 0 & \text{if } y_{i1} < y_{j2} \end{cases},$$

then the test statistic is the sum

$$U^* = \sum_{i=1}^{m} \sum_{j=1}^{n} \delta_{ij}.$$

In the absence of ties (that is, $y_{i1} \neq y_{j2}$ for all i and j) $U^* \sim U(m, n)$ with

$$\mu_U = \frac{mn}{2}, \quad \text{and} \quad \sigma_U^2 = \frac{mn(m + n + 1)}{12}.$$

In this case exact p-values are computed using

$$\text{p-value} = \begin{cases} P(U \leq U^*) & \text{for a left-tailed test; and} \\ \\ P(U \geq U^*) & \text{for a right-tailed test;} \end{cases}$$

and, for a two-tailed test,

$$\text{p-value} = \begin{cases} 2P(U \leq U^*) & \text{if } P(U \leq U^*) \leq 0.5; \text{ and} \\ \\ 2P(U \geq U^*) & \text{if } P(U \leq U^*) > 0.5. \end{cases}$$

The `wilcox.test` function performs the relevant computations associated with the Mann–Whitney U statistic, with relevant argument assignments being

```
wilcox.test(x = y1, y = y2,
    alternative = c("two.sided", "less", "greater"),
    paired = FALSE, exact = NULL, correct = TRUE)
```

where `y1` and `y2` are vectors containing the two independent samples. In the presence of ties `wilcox.test` defaults to the normal approximation with a correction for continuity. Otherwise, in the absence of ties and if $m + n < 50$, the default is to perform the exact test.

Example

Continuing with the `elemStatStudy` data, to illustrate performing the Mann–Whitney test, consider testing a claim that the median score of students less than or equal to 35 years old in project-based courses is lower than the median score of students over 35 years in project-based courses.

For convenience in coding, the desired data are extracted and placed in two vectors, `tradStud` and `nonTradStud`. While code can be prepared to perform the computational tasks described above (see script file). Running

```
> wilcox.test(x = tradStud, y = nonTradStud,
+     paired = FALSE, exact = TRUE, alternative = "less")
```

gives p-value = 0.0521. So, the claim is supported for $\alpha \geq 0.05421$.

Notice that, in the output, even though the statistic is named W instead of U, as stated in the documentation page for this function, the Mann–Whitney approach to the test is conducted.

16.7.2   The Normal Approximation

In [77, p. 699] it is stated that if the sample sizes satisfy $m, n > 8$ (see also, for example, [7, p. 484]) then

$$\frac{U^* - \mu_U}{\sigma_U} \underset{\text{approx}}{\sim} N(0, 1),$$

and if ties are present, in [70, p. 109], it is suggested that the variance be adjusted to

$$\sigma_U^2 = \frac{mn}{12} \left[m + n + 1 - \frac{\sum_{k=1}^g t_k (t_k - 1)(t_k + 1)}{(m+n)(m+n-1)} \right],$$

where t_k represents the number of ties in the kth tied group as described in Section 16.5.2.

To approximate the p-value using $N(0, 1)$, in particular if ties are present in the data, use the strategy applied for the Sign Rank test. That is, the test statistic, including a correction for continuity, is obtained by

$$z^* = \frac{U^* - \mu_U + 0.5\,\delta}{\sigma_U},$$

where

$$\delta = \left\{ \begin{array}{ll} 1 & \text{for a left-tailed probability, and} \\ -1 & \text{for a right-tailed probability.} \end{array} \right.$$

Then the p-value is computed in the same manner as for the previous normal approximation tests.

Example: No Ties Present

Consider implementing the normal approximation, with a correction for continuity, to the problem in the previous example. This is accomplished by running

```
> wilcox.test(x = tradStud, y = nonTradStud,
+     paired = FALSE, exact = FALSE, alternative = "less")
```

to get p-value $= 0.05407$.

Example: Ties Present

It is a simple matter to force ties to appear in the above data by rounding the scores in `tradStud` and `nonTradStud` to the nearest integer. It turns out that the resulting rounded samples contain 79 ties! Consider using these rounded samples (call them `intTradStud` and `intNonTradStud`) to test the claim that there is no difference between the previously identified medians.

Then, running code altered for this scenario (see the script file) it is found that $U^* = 656.5$ and p-value $= 0.1025$. Alternatively, run

```
> wilcox.test(x = intTradStud, y = intNonTradStud,
+       paired = FALSE, exact = FALSE,
+       alternative = "two.sided")
```

to get p-value $= 0.1025$.

16.7.3   The Wilcoxon Rank Sum Test

In this case the two samples are combined and the resulting combined sample of size $m+n$ is sorted in ascending order. Ranks, r_{ij}, are then assigned to each y_{ij}, using average ranks for ties, and the test statistic is obtained by summing the ranks corresponding to terms from the first sample,

$$W^* = \sum_{i=1}^{m} r_{i1},$$

and, in the absence of ties, $W^* \sim W(m, n)$.

As mentioned in Section 16.2.2, regardless of whether ties are present, the statistic W^* is related to a corresponding Mann–Whitney statistic U^* by

$$W^* = U^* + \frac{m(m+1)}{2}.$$

So, the Wilcoxon W-statistic is easily converted to the corresponding Mann–Whitney U-statistic and the methods from Sections 16.7.1 and 16.7.2 along with the function `pwilcox` apply. See the script file for sample code used on the three previous examples.

16.8   Using the Kruskal–Wallis Test to Test Medians

For purposes of this section the data have the form of those used in the median test of Section 16.4. Before beginning, it is useful to bring up some comments from the literature on how the Kruskal–Wallis test can be applied to comparing $J > 2$ medians.

A helpful discussion of the uses (and misuses) of this test appears in [89, pp. 158–161].[8] In discussing the null hypothesis of this test, McDonald begins by stating this as: "The mean ranks of the groups are the same." Another commonly stated null hypothesis (see, for example, [27, p. 290]) is equivalent to: "The population distributions are the same." Of interest for this section is the hypothesis that "The population medians are the same" (see also, for example, [123, pp. 596]). For purposes of testing this hypothesis, McDonald emphasizes the need for the assumption that the underlying random variables have distributions of the same shape.

So, for independent random samples obtained from $J > 2$ populations having continuous underlying random variables with distributions of the same shape, the Kruskal–Wallis test may be used to test the same hypotheses as for the median test, that is,

$$H_0 : \tilde{\mu}_1 = \tilde{\mu}_2 = \cdots = \tilde{\mu}_J \quad \text{vs.} \quad H_1: \text{The medians are not all equal.}$$

Here is an outline of the computational steps involved.

For $j = 1, 2, \ldots, J$ and $i = 1, 2, \ldots, n_j$, let y_{ij} denote i^{th} observation in the jth sample.[9] The J samples are combined, arranged in ascending order and assigned ranks, r_{ij}, with tied cases being assigned average ranks. Next, rank sums, R_j, for each sample are obtained, that is, for each $j = 1, 2, \ldots, J$, compute

$$R_j = \sum_{i=1}^{n_j} r_{ij}.$$

Finally, with $n = \sum_{i=1}^{n_j} n_j$, the Kruskal–Wallis test statistic is given by

$$H^* = \frac{12}{n(n+1)} \sum_{j=1}^{J} \left(R_j^2 / n_j \right) - 3(n+1).$$

Under the null hypothesis and, for $j = 1, 2, \ldots, J$, if either $J = 3$ with $n_j \geq 6$, or if $J > 3$ with $n_j > 5$, the test statistic H approximates the χ^2-distributed with $\nu = p - 1$ degrees of freedom (see, for example, [77, p. 707]).

In the presence of ties in the y_{ij}, it is recommended that H^* be adjusted using

$$H^* = \frac{\frac{12}{n(n+1)} \sum_{j=1}^{J} \left(R_j^2 / n_j \right) - 3(n+1)}{1 - \sum_{k=1}^{g} (t_k^3 - t_k)/(n^3 - n)},$$

---

[8]This citation is for the print version, a link to which McDonald provides with the web version of his book at http://www.biostathandbook.com/index.html

[9]This is equivalent to identifying each case as a pair (y_{ij}, g_j) from a data frame containing a vector y of the observed responses, and a factor g that contains the corresponding sample identifier for each observed response in y.

where t_k represents the number of ties in the kth tied group as described in Section 16.5.2. Being an upper-tailed test, the null hypothesis is rejected if p-value $= P(\chi^2 \geq H^*) \leq \alpha$.

The function `kruskal.test` performs the necessary computations for this test, and it can be called in one of two forms,

```
kruskal.test(formula = y ~ g, data, subset, na.action, ...)
kruskal.test(x, g, ...)
```

where x and y are assigned the response vector and g the factor (or grouping variable). The arguments `data` and `subset` serve the same purposes as previously described, and `na.action` defaults to `"na.omit"`; rows with NA entries are omitted.

Example

For purposes of illustration, consider applying the Kruskal–Wallis test to the example from Section 16.4.1. That is, with $\tilde{\mu}_{\text{Mat}}$, $\tilde{\mu}_{\text{Nat}}$ and $\tilde{\mu}_{\text{Soc}}$ denoting the true median scores of students from the mathematical sciences, natural sciences and social sciences, respectively, in traditionally taught courses, test the hypotheses

$$H_0 : \tilde{\mu}_{\text{Mat}} = \tilde{\mu}_{\text{Nat}} = \tilde{\mu}_{\text{Soc}} \qquad \text{vs.} \qquad H_1: \text{The medians are not all equal.}$$

After extracting the relevant data (and placing these in a data frame `sub`), a quick examination of density histograms suggests that the shapes of the three underlying distributions are fairly similar, so it might be reasonable to use this test. Then,

```
> with(data = sub,
+     expr = kruskal.test(x = score, g = field))
```

gives p-value $= 0.2977$. This result is quite different from the median test performed in Section 16.4.1, evidence of statistically significant differences between median scores is not present here. This being despite the fact that the sample median of the social sciences students is quite a bit larger than those of the other two groups.

One might suspect that what may really have been performed was a test of the hypotheses

$$H_0: \text{The distributions are the same} \qquad \text{vs.}$$
$$H_1: \text{The distributions are not the same.}$$

Quick checks (see script file) reveal that there is insufficient evidence to reject both the normality and constant variances assumptions on the underlying random variables. Moreover, it is the case that an assumption of equal means is also not rejected. So, it is reasonable to not reject the null hypothesis that the distributions are the same.

This re-enforces the cautions raised by McDonald in [89, pp. 158–161] about using the Kruskal–Wallis test. In fact, there it is recommended that the only time this test should be used is when the original data are associated with one ordinal/ranked variable and one nominal variable.

16.9   Working with Ordinal Data

Since data measured using the ordinal level of measurement can be ranked, the rank-based tests from Sections 16.5 through 16.8 can be applied to ordinal data. In such cases, for the most part, the only expectation on the data is that they come from a symmetric distribution and are obtained randomly. In fact, one might choose to reserve the use of rank-based tests for ordinal data and prefer other methods for interval or ratio level data. However, a review of the literature suggests that this is not always necessary. For this section, concerns about the most appropriate method are put aside and the focus is on how to apply the rank based tests from previous sections to ordinal data.

The sample space for an experiment involving ordinal data comprises a discrete (and finite) collection of possible outcomes whose values are ranked from lowest to highest, for example, $\{a_1, a_2, \ldots, a_k\}$ in which a_1 is the outcome with lowest value and a_k has the highest value. The values assigned to the outcomes are typically represented by the positive integer marking the position of a given outcome relative to the others in the sample space. So, while a meaningful (and exact) distance between two outcomes (say a_3 and a_4) cannot be numerically quantified in terms of a well defined unit of measure, it is meaningful to say that a_4 is ranked higher than a_3. This assignment of relative values to each outcome is what permits the use of rank based tests on ordinal data.

One point to note about using rank based tests on ordinal data is that ties will be present in the data, sometimes in large numbers. This is to be expected since samples will contain repetitions of one or more candidate outcomes from the sample space. This means that normal approximations as opposed to exact tests end up being conducted.

16.9.1   Paired Samples

As described in Section 16.6, for $j = 1, 2$, let $y_{1j}, y_{2j}, \ldots, y_{Nj}$ be two dependent samples associated with the same N subjects under two different experimental treatments. But, here consider the case where the underlying random variables have an ordinal level of measurement, and that the observations in these samples come from the same ranked sample space, say $\{a_1, a_2, \ldots, a_k\}$.

Assume that the elements of the sample space are ranked from lowest to highest. Then, integer values from 1 through k corresponding to each of the observations in the samples $y_{1j}, y_{2j}, \ldots, y_{Nj}$ are identified, and differences in values for each pair y_{i1} and y_{i2} are obtained. For example, if y_{i1} corresponds to a_7, then y_{i1} is considered to have a value of 7 and if y_{i2} corresponds to a_3, then y_{i2} is considered to have a value of 3, giving $d_i = y_{i1} - y_{i2} = 4$. Once this is done, the process is exactly as described in Section 16.6, and demonstrated in the second example in Section 16.6.

The following example involves data of the form used in [39] for the purpose of comparing the application of various tests on such data.

Example: Wilcoxon Signed Rank Test

The instructor of a general education quantitative literacy course wished to perform an informal comparison of student opinions about a flipped version of her classes against opinions about traditional lecture-based classes. For two sections of her Fall of 2016 college algebra class (one offered during the day and the other at night), she began the semester using traditional lectures up to the first midterm examination. The last page of the midterm contained an item prompting students for their opinions of this first part of the course. This item contained the statement "Overall, I enjoyed how the content for the course was presented," and students were asked to select one choice from the response options: Strongly Disagree; Disagree; Neutral; Agree; Strongly Agree.[10]

She then switched to a flipped setting, delivering content using ideas she learned about at a short course on teaching flipped classes. She continued this until the second midterm examination, the last page of which contained the same item prompting responses addressing the second (flipped) part of the course.

She used these paired data to determine if there was a statistically significant difference in student opinions about the delivery method for the two parts of the course, the hypotheses being

H_0: Opinions do not differ   vs. H_1: Opinions do differ.

The data for the last item in the survey are contained in the data frame `flipStudy` under the column names `trad` and `flip`.

The response options available to the students can be considered ordinal, ranked from lowest (Strongly Disagree—1) to highest (Strongly Agree—5) in value. So, in testing for a change in opinions, the integer values associated with the levels of `trad` and `flip` are used. A quick examination of the frequency distribution of the integer scores suggests that the distribution of

---

[10]Survey items such as these are referred to as 5-point Likert items. See the original article by Likert, [83], and other later works such as, for example, [84, Ch. I], [21], [39], and [91] for more on the use of Likert items and tests associated with data obtained from Likert items (and scales).

the differences is approximately symmetric and so to test the hypotheses, the `wilcox.test` function is run using, for example,

```
> with(data = flipStudy,
+       expr = wilcox.test(x = as.integer(trad),
+           y = as.integer(flip), paired = TRUE,
+           alternative = "two.sided", exact = FALSE))
```

to get p-value = 0.01237. So, there is a statistically significant difference in opinions about the two delivery methods for any $\alpha \geq 0.01237$.

Basic code to perform the same computations is given in the script for this section. Running that code shows that $\mu_{T+} = 564$. Since the test statistic, $T^* = 331$, is less than μ_{T+} and the null hypothesis is rejected the evidence suggests that students had a higher opinion of the flipped approach to teaching.

What If the Number of Zero Differences Is Large?

This can happen, and since tied observations are excluded in the computations the number of remaining observations can become small. A remedy for such situations was put forward by Pratt in [105]. There, zero-differences are included in the ordering and ranking process of the absolute differences before obtaining the rank sum. In [39] this is referred to as *Pratt's test*, and this was found to be the preferred approach for small paired samples in which the correlation between the paired groups is strong; See the script for this section for sample code.

16.9.2   Two Independent Samples

As for the paired-samples test, ordinal data for two independent samples are assigned integer values according to some criteria. Once this is done, the process is as described in Section 16.7.2.

Example: The Mann–Whitney Test

Using the previously described `flipStudy` data, consider determining if students from the night sections are more likely to respond favorably with respect to the flipped part of the class as compared to students from the day time section. The hypotheses in this case are

$$H_0 \quad : \quad \text{Opinions do not differ between sections} \quad \text{vs.}$$
$$H_1 \quad : \quad \text{Night class students prefer flipped classes.}$$

Since the levels of the factor `section`, also contained in `flipStudy`, are in the order `day` then `night`, the alternative hypothesis states that the opinions of night class students are ranked higher. Thus, this is a left-tailed test and so run

```
> with(data = flipStudy,
+     expr = wilcox.test(formula=as.integer(flip) ~ section,
+         paired = FALSE, alternative = "less",
+         exact = FALSE))
```

to get p-value $= 0.7283$. This suggests that there is insufficient evidence to support the alternative hypothesis.

16.9.3 More than Two Independent Samples

As with the previous two rank based tests, the Kruskal–Wallis test may be conducted on ordinal data. Once numeric scores to enable ranking are assigned to the ordered classes, the procedure is as described in Section 16.8.

Example: Kruskal–Wallis Test

For purposes of illustration, the elemStatStudy from previous examples is altered so that instead of scores, each student is assigned a letter grade (from A, B, C, D and F). These are stored under grades in a new data frame called letterGrades; see the script file for relevant code.

Now consider testing the hypotheses

H_0 : Letter grades across fields are identically distributed, vs.

H_1 : Letter grades across fields are not identically distributed

by running

```
> with(data = letterGrades,
+     expr = kruskal.test(x = as.integer(grades), g = field))
```

to get p-value $= 0.959$. Out of curiosity, one could argue that the grade points assigned to letter grades can be thought of as belonging to an interval level variable, and run the usual one-way F-test described in Section 15.3 and Welch's F-test described in 15.5.2 (see the script for this section). It is evident from the literature that there is always a question as to which procedure is most appropriate, or acceptable, when non-numeric data are assigned numeric values using some rule or the other.

16.9.4 Some Comments on the Use of Ordinal Data

A common refrain in the literature concerning the use of ordinal data in statistical studies is to exercise caution in viewing assigned quantitative (ranking) measures as behaving like data associated with interval level variables. Presumably, there is the tendency to assume adjacent (quantitative) levels in ordinal data as being equidistant in the interval sense. However, the absence of a well defined unit of measure can bring into question matters of precision under such an assumption. Needless to say, there are plenty of examples in the literature that question the soundness of using ordinal data to mimic quantitative measures.

This being said, there are those who argue in favor of the use of ordinal data for pilot studies, particularly when the characteristics of interest are either hard or impossible to quantify. Indeed, there are disciplines in which, more often than not, studies are based on the analyses of such data. Additionally, discussions have arisen with respect to which is the preferred method of analysis for a particular task involving ordinal data. See, for example, works such as [21], [46], [66], [101], [120], [129], and [141], and the earlier cited works in this section, for a flavor of the various views on this subject. The take-home message here is that a careful review of the literature for commonly accepted practices should be conducted prior to using ordinal level data of any form in a statistical study.

More details and/or discussions on extensions of the methods from this chapter (and others) can be found in, for example, [27], [70], and [123].

17

Dependence and Independence

This chapter starts with a discussion on assessing the extent to which a bivariate sample comes from a bivariate normal distribution. Then, four approaches to assessing dependencies between two variables are presented. The script file `Chapter17Script.R` contains code for all examples, and more, presented in this chapter.

17.1 Assessing Bivariate Normality

The function `mvrnorm` from package `MASS` [135] is needed for this section. Install this package and load it into the workspace before beginning.

The focus of this section is on using three characteristics, primarily along the lines of [74, pp. 194–200], to assess the soundness of an assumption of bivariate normality. For an in-depth discussion on the theory of bivariate normality see, for example, [60] or [128].

Let X and Y be random variables such that the pair (X, Y) has a bivariate normal distribution with means μ_X and μ_Y, and variance-covariance matrix

$$\boldsymbol{\Sigma} = \begin{bmatrix} \sigma_X^2 & \sigma_{XY} \\ \sigma_{YX} & \sigma_Y^2 \end{bmatrix}.$$

This matrix is symmetric and positive-definite.[1] At the heart of the methods to be presented here are what are referred to as squared generalized distances.

Squared Generalized Distances

It turns out that the centroid of a bivariate normal distribution is (μ_X, μ_Y), and of interest is a way to measure the distance d of any point (x, y) from the centroid. The measure, d, used here is defined by

$$d^2 = (\mathbf{x} - \boldsymbol{\mu})' \, \boldsymbol{\Sigma}^{-1} \, (\mathbf{x} - \boldsymbol{\mu}),$$

where $\boldsymbol{\Sigma}^{-1}$ is the inverse of $\boldsymbol{\Sigma}$, $\mathbf{x}' = (x, y)$, and $\boldsymbol{\mu}' = (\mu_X, \mu_Y)$. It also turns

---

[1] That is, the matrix and its transpose are equal, $\boldsymbol{\Sigma}' = \boldsymbol{\Sigma}$, and for any nonzero vector $\mathbf{v}$, the inequality $\mathbf{v}' \boldsymbol{\Sigma} \mathbf{v} > 0$ holds true.

out that if (x, y) is allowed to vary in the xy-plane for a fixed d^2, this relation defines an ellipse of the form

$$a_{11} (x - \mu_X)^2 + 2a_{12} (x - \mu_X)(y - \mu_Y) + a_{22} (y - \mu_Y)^2 = d^2$$

where the coefficients a_{ij}, $i, j = 1, 2$ are entries of the matrix Σ^{-1}.

Moving to any bivariate sample (x_i, y_i), $i = 1, 2, \ldots, n$, the sample estimate of Σ is given by

$$\mathbf{S} = \frac{1}{n-1} \begin{bmatrix} \sum_{i=1}^n (x_i - \bar{x})^2 & \sum_{i=1}^n (x_i - \bar{x})(y_i - \bar{y}) \\ \sum_{i=1}^n (x_i - \bar{x})(y_i - \bar{y}) & \sum_{i=1}^n (y_i - \bar{y})^2 \end{bmatrix}$$

and the squared generalized distance of (x_i, y_i) from $(\bar{x}, \bar{y})$ is then given by

$$d_i^2 = (\mathbf{x}_i - \bar{\mathbf{x}})' \mathbf{S}^{-1} (\mathbf{x}_i - \bar{\mathbf{x}}),$$

where $\mathbf{S}^{-1}$ is the inverse of $\mathbf{S}$, $\mathbf{x}_i' = (x_i, y_i)$, and $\bar{\mathbf{x}}' = (\bar{x}, \bar{y})$. Computing the vector of squared generalized distances for any given sample with R is fairly quick, see the script file for a demonstration.

The desired property of such squared generalized distances, d^2, of elements contained in a bivariate normal distribution from the centroid of the distribution is $d^2 \sim \chi^2(2)$.

Trends in Scatterplots

The first characteristic of interest is that plotted points in scatterplots of samples from bivariate normal distributions will exhibit an (approximate) elliptical pattern. This behavior can be illustrated with R as follows.

Simulations of bivariate normal random samples can be obtained using the function `mvnorm` from package `MASS`. The usage definition of this function, along with arguments of interest, is

```
mvrnorm(n = 1, mu, Sigma)
```

where `n` is assigned the sample size desired, `mu` is assigned the means of X and Y, and `Sigma` the symmetric positive-definite variance-covariance matrix. Then, a scatterplot of the sample can be constructed, and an ellipse that bounds a (theoretically expected) percentage of the plotted points can be superimposed using the function

```
ellipse.plot(x, y, prob = 0.5, mu = NULL, sigma = NULL, ...)
```

For this function the expected percentage assigned through `prob` defines the size of the ellipse drawn. By default, the means `mu` for the two random variables and the variance-covariance matrix, `sigma`, are obtained from the sample. See the script file for the code and a demonstration.

Gamma Plots

Since, for any bivariate normal distribution $d^2 \sim \chi^2(2)$, it is expected that a chi-square probability QQ-plot (with $\nu = 2$) of the squared generalized distances for any given sample will exhibit an approximate linear trend through the origin, and with slope 1. This is the second characteristic of interest. The function

```
gamma.plot(x1, x2, main, sub,
       ref.line = c("yEqx", "qq", "both", "none"), ...)
```

performs the relevant computations to construct such plots (see, also, Section 8.4.4). The arguments `x1` and `x2` are assigned the samples for X and Y, respectively, and the argument `ref.line` identifies the reference line to be plotted (the default being the line $y = x$). Again, see the script file for a demonstration.

In the lower-right corner of the plots from this demonstration, notation of the form "48% of $d^2 \leq \chi^2_{0.5}(2)$" will be noticed. This brings up the third characteristic of bivariate normal distributions. This is that about 50% of the observations in any bivariate normal sample will have squared generalized distances from the centroid of at most $\chi^2_{0.5}(2)$. For the previous scatterplots, this translates as about 50% of observations in any bivariate normal sample will lie within the ellipse defined by $d^2 \leq \chi^2_{0.5}(2)$.

17.2 Pearson's Correlation Coefficient

In commonly encountered language, the term correlation with regard to two continuous random variables X and Y refers to the existence of a linear relationship between the two variables. For this section the strength and nature of any existing linear relationship is represented by Pearson's correlation coefficient, ρ, which is defined by

$$\rho = \frac{\sigma_{XY}}{\sigma_X \sigma_Y},$$

where σ_{XY} denotes the covariance of X and Y, and σ_X^2 and σ_Y^2 the variances of X and Y, respectively. An estimate of this parameter is obtained as follows.

Let $x_1, x_2, \ldots, x_n$ and $y_1, y_2, \ldots, y_n$ be samples associated with these variables, respectively. Analogous to the population correlation coefficient, the sample correlation coefficient, r, is defined by, and reduces to

$$r = \frac{s_{XY}}{s_X s_Y} = \frac{\sum_{i=1}^{n} (x_i - \bar{x}) (y_i - \bar{y})}{\sqrt{\sum_{i=1}^{n} (x_i - \bar{x})^2 \sum_{i=1}^{n} (y_i - \bar{y})^2}},$$

where s_{XY} denotes the sample covariance of X and Y, and s_X^2 and s_Y^2 the sample variances of X and Y, respectively. In most texts this is typically rewritten in a commonly encountered computational formula.

The function of interest here, with relevant arguments, is

```
cor.test(x, y, method = "pearson",

    alternative = c("two.sided", "less", "greater"),

    exact = NULL, conf.level = 0.95)
```

which computes the desired correlation coefficient and performs the necessary computations for specified tests. Aside from the argument `method`, the arguments and their assignments are as for previously encountered occurrences in hypothesis testing functions. The formula option, similar in application to previously seen test functions, is also available.

Interpreting Values of r

A strong linear relationship between X and Y is characterized by values of $|r|$ that are close to 1 and the absence of a linear relationship is characterized by values of $|r|$ that are close to zero; see the script file for a graphical demonstration of this. However, rules-of-thumb associated with classifying the strength of such linear relationships through interpretations of the magnitude of r over the interval $[-1, 1]$ tend to depend on the discipline of the study in question.[2] For example, in [100, p. 406] (a behavioral sciences text) the author cites [26] in classifying the strength of a linear relationship between two numeric variables through the use of r as follows.

Range of $|r|$	$[0, 0.3)$	$[0.3, 0.5)$	$[0.5, 1]$
Strength	Weak	Medium	Strong

This is followed by a comment that the range for strong correlations is rarely reached in the social sciences. An alternate classification appears in [77, p. 598],

Range of $|r|$	$[0, 0.5]$	$(0.5, 0.0.8]$	$(0.8, 1]$
Strength	Weak	Medium	Strong

So, the take-home message here is that disciplinary conventions concerning Pearson's correlation coefficient should be referred to when using it to perform a correlation analysis of a bivariate sample.

Regardless of the classification of the strength of a linear relationship, the

---

[2]For those interested, the article [108] provides an informative history of the development of this coefficient and thirteen alternative ways to look at it.

sign of r (positive or negative) is determined by the nature of the relationship between the two variables. An increasing trend in the xy-scatterplot results in $r > 0$ indicating what is referred to as a positive correlation, and a decreasing trend produces $r < 0$ which indicates what is referred to as a negative correlation.

17.2.1   An Interval Estimate of ρ

The function `cor.test`, with a two-sided alternative computes an interval estimate at a desired confidence level using the following approach (see, for example, [123, pp. 778–780]). First, Fisher's z-transformation of r to near normality is used to get

$$z_r = \frac{1}{2} \ln \left[\frac{1+r}{1-r} \right],$$

then the interval

$$z_r - z_{\alpha/2} \sqrt{\frac{1}{n-3}} < z_\rho < z_r + z_{\alpha/2} \sqrt{\frac{1}{n-3}}$$

is computed, where $z_{\alpha/2}$ satisfies $P(Z \geq z_{\alpha/2}) = \alpha/2$ with $Z \sim N(0, 1)$. Now, denote the lower and upper bounds of this interval by z_{r_a} and z_{r_b}, respectively. Then a back-transformation of these bounds is performed using

$$r_a = \frac{e^{2z_a} - 1}{e^{2z_a} + 1} \quad \text{and} \quad r_b = \frac{e^{2z_b} - 1}{e^{2z_b} + 1},$$

and the resulting $(1 - \alpha) \times 100\%$ confidence interval of ρ is

$$r_a < \rho < r_b.$$

Sample code for the application of these computational steps, as well as the use of the `cor.test` function, is contained in the script for this sub-section.

17.2.2   Testing the Significance of ρ

The relevant alternative hypotheses are

$$H_1 : \rho < 0; \qquad H_1 : \rho \neq 0; \qquad \text{or} \qquad H_1 : \rho > 0,$$

and a t test of any one of these is conducted in the following manner. It is assumed (and this may need verifying graphically or otherwise) that the underlying random variables X and Y are linearly related and have a bivariate normal distribution. Under these assumptions, and the null hypothesis, the commonly used statistic is

$$t^* = r \sqrt{\frac{n-2}{1-r^2}} \sim t(n-2),$$

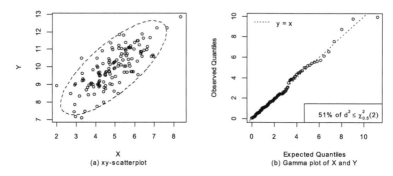

FIGURE 17.1
Plots used to verify the assumptions of linearity and bivariate normality for the data contained in `pearsonData`. See Section 17.1 and script file for construction details. Note that these plots can be supplemented with marginal normal probability QQ-plots of X and Y.

and previously seen conventions for computing p-values involving the t distribution are used (for example, see Section 14.2.3). Here the function `cor.test` with the `method = "pearson"` argument assignment performs the relevant computations.

Example

The dataset `pearsonData` contains a bivariate sample simulated in a manner such that the underlying random variables are approximately linearly related and such that the sample comes from a bivariate normal distribution. This can be verified using the methods from Section 17.1. The plotted points in Figure 17.1(a) follow an approximate linear trend and exhibit an elliptical pattern. Moreover, the gamma plot, Figure 17.1(b), displays the appropriate linear trend and approximately 51% (about 50%) of the squared generalized distances, d_j^2 are less than or equal to $\chi_{0.5}^2(2)$.

So, for the sample in question, the above observations suggest that one may fairly safely assume that the relevant assumptions are (at least approximately) satisfied. Consider testing the hypotheses

$$H_0 : \rho = 0 \qquad \text{vs.} \qquad H_1 : \rho \neq 0.$$

To perform the needed computations for the test, run

```
> with(data = pearsonData,
+       expr = cor.test(x = X, y = Y, method = "pearson",
+             alternative = "two.sided"))
```

to get $r = 0.7672$, $t = 13.899$, $df = 135$, p-value ≈ 0 and the 95% confidence interval $0.688 < \rho < 0.828$. So, a moderate but statistically significant linear relationship between X and Y exists.

17.2.3 Testing a Null Hypothesis with $\rho_0 \neq 0$

Alternatives of the form

$$H_1 : \rho < \rho_0; \qquad H_1 : \rho \neq \rho_0; \qquad \text{or} \qquad H_1 : \rho > \rho_0$$

where $\rho_0 \neq 0$ may also be considered using the approach employed in computing the confidence interval for ρ in Section 17.2.1 (see, for example, [123, pp. 780–781]). In such cases the test statistic

$$z^* = \left(z_r - z_{\rho_0} \right) \sqrt{n - 3} \sim N(0, 1)$$

is used, where Fisher's z-transformation of r to near normality provides

$$z_r = \frac{1}{2} \ln \left[\frac{1 + r}{1 - r} \right] \quad \text{and} \quad z_{\rho_0} = \frac{1}{2} \ln \left[\frac{1 + \rho_0}{1 - \rho_0} \right].$$

The p-values are then computed in the same manner described for the case of $\rho_0 = 0$, but using $Z \sim N(0, 1)$ in this case.[3]

It is a fairly straightforward task to package these computations in a function called, for example,

```
pearsonCor.ztest(x, y, rho,
    alternative = c("less", "two.sided", "greater"))
```

where the argument `rho` is assigned the hypothesized correlation coefficient.

Example

Consider testing the hypotheses

$$H_0 : \rho \leq 0.70 \qquad \text{vs.} \qquad H_1 : \rho > 0.70$$

on the sample contained in `pearsonData` using `pearsonCor.ztest`. Then, running the code

```
> with(data = pearsonData,
+       expr = pearsonCor.ztest(x = X, y = Y, rho = 0.70,
+               alternative = "greater"))
```

gives $z = 1.69305$, p-value $= 0.0452$. So, at $\alpha \geq 0.05$ there is evidence to support the alternative hypothesis that $\rho > 0.70$.

---

[3]Note that while `cor.test` does not perform this procedure, as indicated earlier a confidence interval computed by the function `cor.test` with `method = "pearson"` can be obtained using Fisher's z-transformation (when $n \geq 4$). This confidence interval can then be used to conduct a test of such a hypothesis using an appropriate confidence level. See code in the script for the following example.

17.3   Kendall's Correlation Coefficient

Here too the matter of interest is whether the underlying continuous variables X and Y of a bivariate random sample (x_i, y_i), $i = 1, 2, \ldots, n$ are independent. In this case, however, the only assumption is that the variables X and Y are identically distributed. Notation used to denote the true measure of the strength of the association between X and Y through Kendall's correlation coefficient will be τ, and the corresponding sample correlation coefficient will be denoted by r_K. The following discussion on steps to compute r_K is along the lines of [27, pp. 319–332] and [70, Sec. 8.1–8.3]. See also, for example, [123, Test 30] and [75].

Begin with Q_{ij}, an $n \times n$ matrix, initially with zeros for all its entries. Then, for $1 \le i < j \le n$, redefine

$$Q_{ij} = \begin{cases} 1, & \text{if } (x_j - x_i)(y_j - y_i) > 0, \\ 0, & \text{if } (x_j - x_i)(y_j - y_i) = 0, \\ -1, & \text{if } (x_j - x_i)(y_j - y_i) < 0. \end{cases}$$

Kendall's sample rank correlation coefficient is obtained by computing

$$r_K = \frac{\sum_{i=1}^{n-1} \sum_{j=i+1}^{n} Q_{ij}}{n(n-1)/2}.$$

It should be noted that whether or not ties exist in the data, the above computational formula for Q_{ij} (and r_K) remains the same. The function `cor.test` with the argument assignment `method = "kendall"` applies to this section.

Interpreting Values of r_K

A search of strength classifications for interpreting the value of r_K does not yield the same range of opinions as is the case for r. However, it is reasonable to say, equivalently, that a strong association between X and Y is characterized by values of $|r_K|$ that are close to 1 and a weak association is characterized by values of $|r_K|$ that are close to zero. Moreover, analogous to the case of r, the sign of r_K (positive or negative) is determined by the nature of the relationship between the two variables. An on-average increasing trend in the xy-scatterplot results in $r_K > 0$ indicating a positive correlation, and an on-average decreasing trend produces $r_K < 0$ which indicates a negative correlation.

Aside from the relaxation of the bivariate normality assumption for the previous Pearson's correlation coefficient, r_K can also be used to measure (more legally) the strength of nonlinear monotone (increasing or decreasing) dependencies.

17.3.1 An Interval Estimate of τ

The function `cor.test` does not compute a confidence interval for τ. However, as described in [70, p. 383], a $(1 - \alpha) \times 100\%$ asymptotically distribution-free confidence interval of τ based on r_K can be obtained as follows.

With Q_{ij} and r_K being as defined previously, for each $i = 1, 2, \ldots, n$, first compute

$$C_i = \sum_{j=1}^{n} (Q_{ij} + Q_{ji}) \quad \text{and} \quad \bar{C} = \sum_{i=1}^{n} C_i/n,$$

and define

$$s_K^2 = \frac{2}{n(n-1)} \left[\frac{2(n-2)}{n(n-1)^2} \sum_{i=1}^{n} (C_i - \bar{C})^2 + 1 - r_K^2 \right].$$

Then the $(1 - \alpha) \times 100\%$ confidence interval is given by

$$\max \left\{ -1, r_K - z_{\alpha/2}\, s_K \right\} < \tau < \min \left\{ r_K + z_{\alpha/2}\, s_K, 1 \right\},$$

where $z_{\alpha/2}$ satisfies $P(Z \geq z_{\alpha/2}) = \alpha/2$ with $Z \sim N(0,1)$. These computations are contained in the function

```
kendallZ.interval(x, y, conf.level = 0.95)
```

See sample code for this sub-section as applied to the data `kendallData`.

It is worth mentioning here that `kendallZ.interval` could be used to test hypotheses with $\tau_0 \neq 0$ in a manner analogous to that used for ρ in Section 17.2.3.

17.3.2 Exact Test of the Significance of τ

In the absence of ties and for a suitably small sample size $(n < 50)$, the function `cor.test` with `method = "kendall"` and `exact = TRUE` can be used to conduct the following exact test. The relevant alternative hypotheses are

$$H_1 : \tau < 0; \qquad H_1 : \tau \neq 0; \qquad \text{or} \qquad H_1 : \tau > 0$$

and, under the null hypothesis, the test statistic is

$$T^* = \sum_{i=1}^{n-1} \sum_{j=i+1}^{n} Q_{ij}.$$

Denoting the underlying random variable by T, the p-value is computed using the distribution described in Section 17.7 and the conventions

$$\text{p-value} = \begin{cases} P(T \leq T^*) & \text{for a left-tailed test;} \\ 2\,P(T \geq |T^*|) & \text{for a two-tailed test; and} \\ P(T \geq T^*) & \text{for a right-tailed test.} \end{cases}$$

Observe that $r_K = 2T^*/[n(n-1)]$, and see Section 17.7 for a discussion on the practical limitations that the probability distribution of T^* and r_K places on an exact test.

Example

Consider testing the hypotheses

$$H_0 : \tau = 0; \quad \text{vs} \quad H_1 : \tau \neq 0,$$

on the non-normal bivariate sample in `kendallData`. Running

```
> with(data = kendallData,
+      expr = cor.test(x = X, y = Y,
+          alternative = "two.sided", method = "kendall"))
```

gives $r_K = 0.9371$, $T = 1047$ and p-value ≈ 0.

17.3.3 Approximate Test of the Significance of τ

The exact test of significance of τ cannot be conducted if there are ties in the data or if the sample size is large. However, as stated in [27, p. 327], a normal approximation of τ works well for samples of size $n \geq 8$. These computations are performed by the function `cor.test`, with `method = "kendall"` and `exact = FALSE`. In the presence of ties, or for samples of size $n \geq 50$, the function defaults to this approximate method. The process is as follows.

First obtain the sizes of tied groups in $x_1, x_2, \ldots, x_n$, denote these by $t_1, t_2, \ldots, t_g$ where $g \leq n$ and where untied entries are assigned a tied-group size of 1 (see Section 16.5.2). Similarly, obtain the sizes of tied groups in $y_1, y_2, \ldots, y_n$ and denote these by $u_1, u_2, \ldots, u_h$ where $h \leq n$. Then, under the null hypothesis, the test statistic is computed using

$$z^* = T^*/\sigma_T,$$

where σ_T is obtained using the (somewhat formidable) formula

$$
\begin{aligned}
\sigma_T^2 = &\left\{ n(n-1)(2n+5) - \sum_{i=1}^{g} t_i(t_i-1)(2t_i+5) \right.\\
&\left. - \sum_{j=1}^{h} u_j(u_j-1)(2u_j+5) \right\} \bigg/ 18\\
&+ \frac{\sum_{i=1}^{g} t_i(t_i-1)(t_i-2) \sum_{j=1}^{h} u_j(u_j-1)(u_j-2)}{9n(n-1)(n-2)}\\
&+ \frac{\sum_{i=1}^{g} t_i(t_i-1) \sum_{j=1}^{h} u_j(u_j-1)}{2n(n-1)}.
\end{aligned}
$$

Observe that in the absence of ties this reduces to

$$\sigma_T^2 = \frac{n(n-1)(2n+5)}{18}.$$

Then, p-values are computed using the usual conventions with $Z \sim N(0,1)$.

Example

Consider testing the same hypotheses

$$H_0 : \tau = 0; \qquad \text{vs} \qquad H_1 : \tau \neq 0,$$

on the non-normal bivariate sample in `kendallData` by running

```
> with(data = kendallData,
+      expr = cor.test(x = X, y = Y, exact = FALSE,
+          alternative = "two.sided", method = "kendall"))
```

In this case the output produced is $r_K = 0.9371$, $z^* = 9.2897$ and p-value ≈ 0.

17.4   Spearman's Rank Correlation Coefficient

The method described here provides yet another alternative to Pearson's method. This is a rank-based method, and what is measured here is the strength of the correlation of the ranks of the entries for the two samples rather than the entry values themselves. In effect, Pearson's correlation coefficient on the ranks (as opposed to the entries) is computed. Some denote the true correlation coefficient by ρ_S, and the sample correlation coefficient is typically denoted by r_S. The following is along the lines of [70, Sec. 8.5] with additional comments coming from [123, Test 29].

Consider a bivariate sample (x_i, y_i), $i = 1, 2, \dots, n$ for which either the bivariate normality assumption does not hold, or for which nothing is known about the underlying bivariate distribution. Additionally, the underlying random variables X and Y need not be continuous. They are, however, assumed related in the monotone increasing or decreasing sense.[4] Here is the process.

First obtain the ranks of the observations within each sample separately, assigning mean ranks to ties. Call the resulting ranks R_{xi} and R_{yi}, respectively, and obtain the paired differences $d_i = R_{xi} - R_{yi}$. Next, in the presence of ties, first obtain the sizes of tied groups in $x_1, x_2, \dots, x_n$, denote these by

---

[4]Note that since the ranks of the sample entries are what are considered, this procedure may also be applied to bivariate samples in which the underlying variables are ordinal, for example, in evaluating the degree of agreement between the rankings of two judges for n subjects/objects [123, p. 863].

$t_1, t_2, \ldots, t_g$ where $g \leq n$ and where untied entries are assigned a tied-group size of 1 (see Section 16.5.2). Similarly, obtain the sizes of tied groups in $y_1, y_2, \ldots, y_n$ and denote these by $u_1, u_2, \ldots, u_h$ where $h \leq n$.

Finally, if ties are present in the data, Spearman's rank correlation coefficient is computed using

$$r_S = \frac{n(n^2 - 1) - 6 \sum_{i=1}^{n} d_i^2 - \frac{1}{2} \left\{ \sum_{j=1}^{g} \left[t_j \left(t_j^2 - 1 \right) \right] + \sum_{k=1}^{h} \left[u_k \left(u_k^2 - 1 \right) \right] \right\}}{\sqrt{\left[n(n^2 - 1) - \sum_{j=1}^{g} t_j \left(t_j^2 - 1 \right) \right] \left[n(n^2 - 1) - \sum_{k=1}^{h} u_k \left(u_k^2 - 1 \right) \right]}}.$$

In the absence of ties, the above formula reduces to

$$r_S = 1 - \frac{6 \sum_{i=1}^{n} d_i^2}{n(n^2 - 1)}.$$

The function `cor.test` with `method = "spearman"` applies to this section.

Interpretation of r_S

Strong associations between the variables/ranks are indicated by $|r_S|$ close to 1 and weak associations by $|r_S|$ close to 0. Additionally, since r_S is computed using relative ranks, it can also be used to measure the strength of non-linear monotone increasing (or decreasing) associations between two continuous variables. See, for example, [89, p. 211].

17.4.1   Exact Test of the Significance of ρ_S

The relevant alternative hypotheses may be stated in the form

$$H_1 : \rho_S < 0; \qquad H_1 : \rho_S \neq 0; \qquad \text{or} \qquad H_1 : \rho_S > 0,$$

whereas others state these in a more general manner using, for example, language as follows. If the values of Y tend to increase as the values of X increase, then X and Y are said to be positively associated. Alternatively, if the values of Y tend to decrease as the values of X increase, then X and Y are said to be negatively associated.

An exact test of ρ_S can be conducted if ties are not present in the data and as long as n is not too large. In the absence of ties it is theoretically possible to construct the probability distribution of r_S, see Section 17.7, and in this case r_S is the test statistic. Denote the underlying random variable of r_S by S, then p-values are obtained using

$$\text{p-value} = \begin{cases} P(S \leq r_S) & \text{for a left-tailed test;} \\[2mm] 2\,P(S \geq |r_S|) & \text{for a two-tailed test; and} \\[2mm] P(S \geq r_S) & \text{for a right-tailed test.} \end{cases}$$

This being said, the construction of the probability distribution r_S is not very practical for large samples; see Section 17.7. In fact, in the documentation for the `cor.test` function it is stated that a series approximation is used to obtain exact p-values for Spearman's test for samples of size $n \geq 10$. The function call

```
cor.test(x, y, method = "spearman")
```

performs the exact test by default if there are no ties in the data and the sample size is less than 1290. If the exact test cannot be performed (ties are present), the approximate test is conducted and a warning is printed to the console along with the results of the approximate test. See the script file for sample code of using this procedure on the previous `kendallData`.

17.4.2   Approximate Test of the Significance ρ_S

In the presence of ties an approximate test has to be conducted. Regardless of whether ties are present in the data or not, it can be shown that under the null hypothesis r_S is asymptotically normal (as $n \to \infty$) with mean $\mu_{r_S} = 0$ and variance $\sigma_{r_S}^2 = 1/(n-1)$. In fact, for sufficiently large samples ($n \geq 10$; see Section 17.7) it turns out that r_S is approximately normally distributed and a test of the significance of ρ_S may be conducted using the statistic

$$t^* = \frac{r_S \sqrt{n-2}}{\sqrt{1-r_S^2}} \sim t(n-2),$$

and the p-values are computed using the same conventions as in Section 17.2.2. The earlier function `cor.test` with the arguments `method = "spearman"` and `exact = FALSE` performs the relevant calculations. Again, see the script file for sample code of using this procedure on the previous `kendallData`.

17.5   Correlations in General: Comments and Cautions

It is important to remember that Pearson's correlation coefficient is a measure of the strength of the linear relationship between two continuous variables. This being said, it is also important to remember that samples possessing nonlinear trends can often yield misleading correlation coefficients. For example, run `example(anscombe)` for a demonstration of four different datasets having very different graphical appearances that can yield almost identical sample correlation coefficients. See the source article [4].

It is also important to keep in mind the usual cautions when drawing cause-and-effect conclusions based on relationships observed in the data. See the many (sometimes amusing) examples on the internet for more on this.

17.6   Chi-Square Test of Independence

Consider a study involving two nominal variables, say A with I classes and B with J classes. For $i = 1, 2, \ldots, I$ and $j = 1, 2, \ldots, J$, let o_{ij} represent the frequency with which a particular characteristic is observed under the ith level of the variable A and the jth level of the variable B. Then the data can be organized in a two-way contingency table as in Section 12.1. The chi-square test of independence looks at determining if there is evidence to conclude that the true proportions across classes of one variable (B) depend on which class of the other variable (A) is being considered (or vice versa). The simplest statements of the hypotheses for this test are

$$H_0 : A \text{ and } B \text{ are independent} \quad \text{vs.} \quad H_1 : A \text{ and } B \text{ are not independent.}$$

Assuming the sample is random, the steps for performing the test of independence are similar to previously seen chi-squared tests on nominal data. The process, revisited, is as follows. For $i = 1, 2, \ldots, I$ (rows) and $j = 1, 2, \ldots, J$ (columns), obtain the expected frequencies for each (i, j)-cell using the relation $e_{ij} = r_i\, c_j / n$. Then, if every expected frequency satisfies $e_{ij} \geq 5$, the test statistic

$$\chi^{2*} = \sum_{i=1}^{I} \sum_{j=1}^{J} (o_{ij} - e_{ij})^2 / e_{ij} \sim \chi\left[(I-1)(J-1)\right]$$

is used to compute p-value $= P(\chi^2 > \chi^{2*})$. The earlier seen `chisq.test` function applies to this test, the code being

```
chisq.test(x)
```

where `x` is assigned the $I \times J$ matrix containing the observed frequencies.

Example

A sociologist in a medium-sized California city wishes to test the claim that an unmarried individual's type of lodging (out of condominium, efficiency, house, or tiny house) is dependent on the type of employment (out of construction, government, manufacturing, service, and technology) she/he is engaged in. A random sample of 138 unmarried and employed people are surveyed. The data are contained in the data frame `homeChoice`.

Let A be the type of lodging that a subject lives in, and B the type of employment the same subject is engaged in. Then the hypotheses to be tested are

$$H_0 : A \text{ and } B \text{ are independent} \quad \text{vs} \quad H_1 : A \text{ and } B \text{ are dependent. (Claim).}$$

To perform the relevant computations, run

```
> with(data = homeChoice,

+     expr = chisq.test(x = table(lodging, employment)))
```

to get $\chi^{2*} = 5.7666$, $df = 12$, p-value $= 0.9274$. Also outputted is a warning message that the "chi-squared approximation may be incorrect." This is because the expected frequency for the condominium-construction and condominium-technology cells are less than 5. However, two out of twenty cells (10%) with $e_{ij} < 5$ is below the 20% threshold for chi-square tests involving contingency tables. So, at most traditional levels of significance, there is insufficient evidence to support the claim that an unmarried individual's type of lodging is dependent on the type of employment she/he is engaged in.

17.7   For the Curious: Distributions of r_K and r_S

These notes relate to the exact tests discussed in Section 17.2 and 17.3 and are along the lines of the presentations in [70, pp. 370–371, 398–400]. In the absence of ties and under the null hypothesis that the variables X and Y are mutually independent, the probability distributions of r_K and r_S can be obtained as follows.

For $i, j = 1, 2, \ldots, n$, denote the ranks of x_i and y_i by R_{x_i} and R_{y_i}, respectively. Then, without loss of generality, assume $R_{x_i} = i$; note that this can be arranged by sorting the bivariate pairs (x_i, y_i) in ascending order of the x_i. At this point it is possible to consider all possible rank configurations R_y, of which there are $n!$ equally likely arrangements. Since each rank configuration R_y is equally likely, the probability that a particular configuration will occur is $1/n!$. The task then is to connect each possible value of r_K or r_S (in the absence of ties) to a corresponding probability.

The main task in finding these probabilities lies in generating all possible permutations of the rank configurations R_y. Code for this purpose was adapted from code in the function **permn** in package **combinat** [24].[5] For smaller sample sizes of $n < 10$ (even up to 10) it is possible to generate all permutations of $1, 2, \ldots, n$ fairly quickly, find the associated (Kendall's or Spearman's) correlation coefficients, and then obtain the corresponding probabilities. However, $n!$ becomes very large as n increases in value and the number of iterations needed to find all possible permutations of the numbers $1, 2, \ldots, n$ becomes prohibitive.[6] The script for this section contains sample code to implement

---

[5]The permutation generating algorithm used in **permn**, and adapted for purposes of the notes in this section, is described in [107, p. 170].

[6]For example, $10! = 3628800$, $11! = 39916800$ and $12! = 479001600$. In fact, if $n \geq 10$ the function **cor.test** employs a series approximation to calculate "exact" probabilities associated with r_S.

these ideas and produce density plots specific to any one of the two correlation coefficients using the following layouts.

Kendall's r_K

For convenience, let K denote the underlying random variable for Kendall's rank correlation coefficient r_K. Let Q_{ij} be initialized with zero entries as described in Section 17.3. Then, for each rank configuration of R_y and for $1 \leq i < j \leq n$, compute

$$
Q_{ij} = \left\{ \begin{array}{ll} 1, & \text{if } \left(R_{x_j} - R_{x_i}\right)\left(R_{y_j} - R_{y_i}\right) > 0, \\ -1, & \text{if } \left(R_{x_j} - R_{x_i}\right)\left(R_{y_j} - R_{y_i}\right) < 0 \end{array} \right.
$$

and then compute

$$
r_K = \frac{\sum_{i=1}^{n-1} \sum_{j=i+1}^{n} Q_{ij}}{n\left(n-1\right)/2}.
$$

Next, suppose a given value of r_K occurs for $m \geq 1$ different rank configurations. Then, K being a discrete random variable,

$$
P(K = r_K) = \frac{m}{n!}.
$$

The function `kendalls.dist` computes an exact probability distribution for r_K up to $n = 9$, but runs into trouble (time-wise) for larger sample sizes.

Spearman's r_S

Let W denote the underlying random variable for Spearman's rank correlation coefficient r_S. Since

$$
r_S = 1 - \frac{6\sum_{i=1}^{n} d_i^2}{n(n^2 - 1)}
$$

depends on the paired differences $d_i = R_{x_i} - R_{y_i}$, it is possible for $\sum_{i=1}^{n} d_i^2$ to have the same value for different rank configurations of R_y. Suppose a given value of r_S occurs for $m \geq 1$ different rank configurations. Then, W being a discrete random variable,

$$
P(W = r_S) = \frac{m}{n!}.
$$

The function `spearman.dist` computes an exact probability distribution for r_K up to $n = 9$, but also runs into trouble (time-wise) for larger sample sizes.

Bibliography

[1] Adler, D., C. Gläser, O. Nenadic, J. Oehlschlägel and W. Zucchini. *ff: memory-efficient storage of large data on disk and fast access functions.* R package version 2.2-14, 2018. http://CRAN.R-project.org/package=ff.

[2] Agresti, A., *Categorical Data Analysis*, 2nd ed., John Wiley & Sons, New York, 2002.

[3] Anscombe, F. J., Rejection of Outliers, *Technometrics*, 2, 123–147, 1960.

[4] Anscombe, F. J., Graphs in Statistical Analysis, *The American Statistician*, Vol. 27, No. 1, 17–21, 1973.

[5] Anscombe, F. J. and W. J. Glynn, Distribution of the Kurtosis Statistic b_2 for Normal Samples, *Biometrika*, Vol. 70, No. 1, 227–234, 1983.

[6] Bååth, R., The State of Naming Conventions in R. *The R Journal*, Vol. 4/2, December 2012. http://journal.r-project.org/archive/2012-2/.

[7] Bain, L. J. and M. Engelhardt, *Introduction to Probability and Mathematical Statistics*, 2nd ed., Duxbury Classic Series, Duxbury Press, CA, 2000.

[8] Barnett, V. and T. Lewis, *Outliers in Statistical Data*, 3rd ed., John Wiley & Sons, New York, 1994.

[9] Bartlett, M. S., Properties of Sufficiency and Statistical Tests, *Proceedings of the Royal Society of London. Series A. Mathematical and Physical Sciences*, 160, No. 901, 268–282, 1937.

[10] Belia, S., F. Fidler, J. Williams and G. Cumming, Researchers Misunderstand Confidence Intervals and Standard Error Bars, *Psychological Methods*, 10 No. 4, 389–396, 2005.

[11] Benjamini, Y. and H. Braun, John Tukey's Contributions to Multiple Comparisons, *ETS Research Report Series*, 2002, 2, i–27. doi:10.1002/j.2333-8504.2002.tb01891.x

[12] Bluman, A. G., *Elementary Statistics: A Step by Step Approach*, A Brief Version, 6th ed., McGraw Hill, New York, 2013.

[13] Bonett, D. G. and T. A. Wright, Sample Size Requirements for Estimating Pearson, Kendal and Spearman Correlations, *Psychometrika*, Vol. 65, No. 1, 23–28, 2000.

[14] Box, G. E. P., W. G. Hunter and J. S. Hunter, *Statistics for Experimenters*, Wiley, New York, 1978.

[15] Braun, W. J., *MPV: Data Sets from Montgomery, Peck and Vining's Book*. R package version 1.53, 2018. http://cran.r-project.org/package=MPV.

[16] Bretz, F, T. Hothorn and P. Westfall, *Multiple Comparison Using R*, CRC Press, Boca Raton, FL, 2011.

[17] Broch, C. S. and D. F. Ferreira, *nCDunnett: Noncentral Dunnett's Test Distribution*. R Package version 1.1.0, 2015. https://CRAN.R-project.org/package=nCDunnett.

[18] Brown, M. B. and A. B. Forsythe, Robust Tests for the Equality of Variances, *Journal of the American Statistical Association*, 69, 364–367, 1974.

[19] Brown, M. B. and A. B. Forsythe, The ANOVA and Multiple Comparisons for Data with Heterogeneous Variances, *Biometrics*, 30, 719–724, 1974.

[20] Brown, M. B. and A. B. Forsythe, The Small Sample Behavior of Some Statistics Which Test the Equality of Several Means, *Technometrics*, 16, 129–132, 1974.

[21] Carifio, J. and R. J. Perla, Ten Common Misunderstandings, Misconceptions, Persistent Myths and Urban Legends about Likert Scales and Likert Response Formats and Their Antidotes, *Journal of Social Sciences*, 3, No. 3, 106–116, 2007.

[22] Casella, G. and R. L. Berger, *Statistical Inference*, Wadsworth & Brooks/Cole, Pacific Grove, CA, 1990.

[23] Chambers, J. M., W. S. Cleveland, B. Kleiner, and P. A. Tukey, *Graphical Methods for Data Analysis*, Duxbury Press, Belmont, CA, 1983.

[24] Chasalow, S., *combinat: combinatorics utilities*. R Package version 0.0-8, 2012. https://CRAN.R-project.org/package=combinat.

[25] Cohen, J., An Alternative to Marascuilo's "Large-Sample Multiple Comparisons" for Proportions, *Psychological Bulletin*, 67, No. 3, 199–201, 1967.

[26] Cohen, J., *Statistical Power Analysis for the Behavioral Sciences*, 2nd ed. Lawrence Erlbaum Associates, New Jersey, 2008.

[27] Conover, W. J., *Practical Nonparametric Statistics*, 3rd ed., John Wiley & Sons, New York, 1999.

[28] Conover, W. J., M. E. Johnson and M. M. Johnson, A Comparitive Study of Tests for Homogeniety of Variances, with Applications to Outer Continental Shelf Bidding Data, *Technometrics*, 23, No. 4, 351–361, 1981.

[29] Cowpertwait, P. S. P. and A. V. Metcalfe, *Introductory Time Series with R*, Springer, New York, 2009.

[30] Cryer, J. D. and K. Chan, *Time Series Analysis: With Applications in R*, 2nd ed., Springer, New York, 2008.

[31] Cumming, G., Inference by Eye: Pictures of Confidence Intervals and Thinking About Levels of Confidence. *Teaching Statistics*, 29, No. 3, 2007.

[32] Cumming, G, F. Fidler and D. L. Vaux, Error Bars in Experimental Biology. *The Journal of Cell Biology*, 177, No. 1, 7–11, 2007.

[33] Cumming, G. and S. Finch, Inference by Eye: Confidence Intervals and How to Read Pictures of Data. *American Psychologist*, 60, No. 2, 170–180, 2005.

[34] D'Agostino, R. B., Transformation to Normality of the Null Distribution of g_1, *Biometrika*, Vol. 57, No. 3, 679–681, 1970.

[35] Dalgaard, P., *Introductory Statistics with R*, Springer, New York, 2002.

[36] Day, R. W. and G. P. Quinn, Comparisons of Treatments After an Analysis of Variance in Ecology, *Ecological Monographs*, 59, No. 4, 433–463, 1989.

[37] Dean, A. and D. Voss, *Design and Analysis of Experiments*, Springer, New York, 1999.

[38] DeCarlo, L. T., On the Meaning and Use of Kurtosis, *Psychological Methods*, 2(3), 292–307, 1997.

[39] Derrick, B. and P. White, Comparing Two Samples From an Individual Likert Question, *International Journal of Mathematics and Statistics*, Vol. 18, No. 3, 1–13, 2017.

[40] De Veaux, R. D., P. F. Velleman and D. E. Bock, *Stats: Data and Models*, 3rd ed., Addison-Wesley, Boston, 2012.

[41] Dobson, A. J., *An Introduction to Statistical Modelling*, Chapman and Hall, London, 1983.

[42] Draper, N. R. and W. G. Hunter, Transformations: Some Examples Revisited, *Technometrics*, 11, No. 1, 23–40, 1969.

[43] Dunnett, C. W., A Multiple Comparison Procedure for Comparing Several Treatments with a Control, *Journal of the American Statistical Association*, 50, No. 272, 1096–1121, 1955.

[44] Dunnett, C. W., Pairwise Multiple Comparison in the Unequal Variance Case, *Journal of the American Statistical Association*, 75, No. 372, 796–800, 1980.

[45] Dunnett, C. W. and M. Sobel, A Bivariate Generalization of Student's *t*-distribution, with Tables for Certain Special Cases, *Biometrika*, 42, 258–260, 1955.

[46] Edmondson, D. R., Likert Scales: A History, *CHARM Proceedings*, Vol. 12, 127–133, 2005.

[47] Faraway, J. J., *Linear Models with R*, Chapman & Hall/CRC, Boca Raton, FL, 2005.

[48] Faraway, J. J., *Extending the Linear Model with R: Generalized Linear, Mixed Effects and Nonparametric Regression Models*, Chapman & Hall/CRC, Boca Raton, FL, 2006.

[49] Feder, P. I., Graphical Techniques in Statistical Analysis—Tools for Extracting Information from Data, *Technometrics*, 16, No. 2, 287–299, 1974.

[50] Feinerer, I., An Introduction to Text Mining in R. *R News*, 8(2), 19–22, 2008. http://CRAN.R-project.org/doc/Rnews/

[51] Feinerer, I. and K. Hornik. *tm: Text Mining Package*. R package version 0.7-3, 2017. http://CRAN.R-project.org/package=tm

[52] Feinerer, I., K. Hornik and D. Meyer. Text Mining Infrastructure in R. *Journal of Statistical Software* 25(5): 1–54, 2008. http://www.jstatsoft.org/v25/i05/

[53] Filliben, J. J., The Probability Plot Correlation Coefficient Test for Normality, *Technometrics*, 17, No. 1, 111–117, 1975.

[54] Fligner, M. A. and T. J. Killeen, Distribution-Free Two-Sample Tests for Scale, *Journal of the American Statistical Association*, 71, No. 353, 210–213, 1976.

[55] Fox, J., *Applied Regression Analysis and Generalized Linear Models*, 2nd ed., Sage Publications, Inc., Thousand Oaks, CA, 2008.

[56] Fox, J., *car: Companion to Applied Regression*. R package version 3.0-0, 2018. http://CRAN.R-project.org/package=car.

[57] Gabriel, K. R., A Simple Method of Multiple Comparisons of Means, *Journal of the American Association*, 73, 210–213, 1976.

[58] Garrett, R. G. *rgr: The GSC Applied Geochemistry EDA Package*. R package version 1.1.15, 2018. http://CRAN.R-project.org/package=rgr

[59] Gastwirth, J. L., Y. R. Gel and W. Miao, The Impact of Levene's Test of Equality of Variances on Statistical Theory and Practice, *Statistical Science*, 24, No. 3, 343–360, 2009.

[60] Gnanadesikan, R., *Methods for Statistical Data Analysis of Multivariate Observations*, John Wiley & Sons, New York, 1977.

[61] Genz, A., F. Bretz, T. Miwa, X. Mi, F. Leisch, F. Scheipl, and T. Hothorn, *mvtnorm: Multivariate Normal and t Distributions*. R package version 1.0-8, 2018. http://CRAN.R-project.org/package=mvtnorm.

[62] Good, I. J., What Are Degrees of Freedom? *American Statistician*, 27, No. 2, 227–228, 1973.

[63] Hand, D. J., F. Daly, A. D. Lunn, K. J. McConway and E. Ostrowski, *A Handbook of Small Data Sets*, Chapman & Hall/CRC London, 1994.

[64] Hauck, W. W., A Comparitive Study of Conditional Maximum Likelihood Estimation of a Common Odds Ratio. *Biometrics*, 40, No. 4, 1117–1123, 1984.

[65] Hay-Jahans, C., *An R Companion to Linear Statistical Models*, CRC Press, Boca Raton FL, 2011.

[66] Harwell, M. R. and G. G. Gatti, Rescaling Ordinal Data to Interval Data in Educational Research, *Review of Educational Research*, Vol. 71, No. 1, 105–131, 2001.

[67] Hoaglin, D. C., John W. Tukey and Data Analysis, *Statistical Science*, Vol. 18, No. 3, 311–318, 2003.

[68] Hoaglin, D. C., F. Mosteller, and J. W. Tukey, *Understanding Robust and Exploratory Data Analysis*, John Wiley & Sons, New York, 1983.

[69] Hogg, R. V. and A. T. Craig, *Introduction to Mathematical Statistics*, 4th ed., Macmillan, New York, 1978.

[70] Hollander, M. and D. A. Wolfe, *Nonparametric Statistical Methods*, 2nd ed., John Wiley and Sons, New York, 1999.

[71] Hothorn, T., F. Bretz and P. Westfall, Simultaneous Inferences in General Parametric Models. *Biometrical Journal*, 50(3), 346–363, 2008.

[72] Hotelling, H., The Generalization of Student's Ratio, *Annals of Mathematical Statistics*, 2(3), 360–378, 1931.

[73] Joanes, D. N. and C. A. Gill, Comparing Measures of Sample Skewness and Kurtosis, *Journal of the Royal Statistical Sociery. Series D (The Statistician)*, 47(1), 183–189, 1998.

[74] Johnson, R. A. and D. W. Wichern, *Applied Multivariate Statistical Analysis*, 4th ed., Prentice Hall, Inc., New Jersey, 1998.

[75] Kendall, M. G., A New Measure of Rank Correlation, *Biometrika*, Vol. 30, No. 1/2, 81–93, 1938.

[76] Kimmel, H. D., Three Criteria for the Use of One-Tailed Tests, *Psychological Bulletin*, 54, No. 4, 351–353, 1957.

[77] Kokoska, S., *Introductory Statistics: A Problem-Solving Approach*, W. H. Freeman and Company, New York, 2011.

[78] Komsta, L. and F. Novomestky, *moments: Moments, Cumulants, Skewness, Kurtosis and Related Tests.* R package version 0.14, 2015. http://CRAN.R-project.org/package=moments

[79] Kramer, C. Y., Extension of Multiple Range Tests to Group Means with Unequal Numbers of Replications, *Biometrics*, 12, 302–310, 1956.

[80] Kutner, M. H., C. J. Nachtsheim, J. Neter, and W. Li, *Applied Linear Statistical Models*, 5th ed., McGraw-Hill, New York, 2005.

[81] Levene, H., Robust Tests for the Equality of Variances, in *Contributions to Probability and Statistics: Essays in Honor of Harold Hotelling*, ed. I. Olkin, Stanford University Press, 278–292, Palo Alto, CA, 1960.

[82] Li, C. C., *Introduction to Experimental Statistics.* McGraw Hill, New York, 1964.

[83] Likert, R., A Technique for the Measurement of Attitudes, *Archives of Psychology*, Vol. 22, No. 140, 5–55, 1932.

[84] Likert, R. A. and S. P. Hayes, Jr. (Editors), *Some Applications of Behavioural Research*, Paris, UNESCO, 1957.

[85] Loftus, G. R. and M. E. Masson, Using Confidence Intervals in Within-Subject Designs. *Psychonomic Bulletin & Review*, 1, No. 4, 476–490, 1994.

[86] Looney, S. W. and T. R. Gulledge Jr., Use of the Correlation Coefficient with Normal Probability Plots, *The American Statistician*, 39, No. 1, 75–79, 1985.

[87] Marascuilo, L. A., Large-Sample Multiple Comparisons, *Psychological Bulletin*, 65, No. 5, 280–290, 1966.

[88] Masson, M. E. and G. R. Loftus, Using Confidence Intervals for Graphically Based Data Interpretation. *Canadian Journal of Experimental Psychology*, 57, No. 3, 203–220, 2003.

[89] McDonald, J. H., *Handbook of Biological Statistics*, 3rd ed., Sparky House Publishing, Baltimore, MD, 2014.

[90] McGill, R., J. W. Tukey, and W. A. Larsen, Variations of Box Plots, *The American Statistician*, 32(1), 12–16, 1978.

[91] Meek, G. E., C. Ozgur, and K. Dunning, Comparison of the t vs. Wilcoxon Signed-Rank Test for Likert Scale Data and Small Samples. *Journal of Modern Applied Statistical Methods*, Vol. 6, No. 1, Art. 10, 91–106, 2007.

[92] Moder, K., How to Keep the Type I Error Rate in ANOVA if Variances are Heteroscedastic. *Austrian Journal of Statistics*, 36(3), 179–188, 2007.

[93] Moder, K., Alternatives to F-Test in One-Way ANOVA in Case of Heterogeniety of Variances (A Simulation Study), *Psychological Test and Assessment*, 52(4), 343–353, 2010.

[94] Montgomery, D. C., *Design and Analysis of Experiments*, 4th ed., John Wiley & Sons, New York, 1997.

[95] Moore, D., *The Basic Practice of Statistics*, 5th ed., W. H. Freeman, New York, 2009.

[96] Murrell, P., *R Graphics*, 2nd ed., The R Series, CRC Press, Boca Raton, FL, 2012.

[97] Newcombe, R. G., Two-Sided Confidence Intervals for the Single Proportion: Comparison of Seven Methods. *Statistics in Medicine* 17, 857–872, 1998.

[98] Newcombe R.G., Interval Estimation for the Difference Between Independent Proportions: Comparison of Eleven Methods. *Statistics in Medicine* 17, 873–890, 1998.

[99] Newman, D., The Distribution of Range in Samples from a Normal Population, Expressed in Terms of an Independent Estimate of Standard Deviation, *Biometrika*, 31, No. 1/2, 20–30, 1939.

[100] Nolan, S. A. and T. E. Heinzen, *Statistics for the Behavioral Sciences*, 2nd ed., Worth Publishers, New York, 2012.

[101] O'Brien, R. M., Using Rank Category Variables to Represent Continuous Variables: Defects of Common Practice, *Social Forces*, Vol. 59, No. 4, Special Issue, 1149–1162, 1981.

[102] Pandey, S and C. L. Bright. What Are Degrees of Freedom, *Social Work Research*, 32, No. 2, 119–128, 2008.

[103] Pillemer, D. B., One- versus Two-tailed Hypothesis Tests in Contemporary Educational Research, *Educational Researcher*, 20, No. 9, 13–17, 1991.

[104] Pillia, K. C. S., On the distribution of the 'Studentized' Range, *Biometrika*, 39, No. 1/2, 194–195, 1952.

[105] Pratt, J. W., Remarks on Zeros and Ties in the Wilcoxon Signed Rank Procedures. *Journal of the American Statistical Association*, 54, No. 287, 655–667, 1959.

[106] Rafter, J. A., M. L. Abell and J. P. Braselton. Multiple Comparison Methods for Means. *SIAM Review*, 44, No. 2, 259–278, 2002.

[107] Reingold, E. M., J. Nievergelt, and N. Deo. *Combinatorial Algorithms: Theory and Practice*, Prentice Hall, New Jersey, 1977.

[108] Rodgers, L. J. and W. A. Nicewander, Thirteen Ways to Look at the Correlation Coefficient. *The American Statistician*, Vol. 42, No. 1, 59–66, 1988.

[109] Royston, J. P., An Extension of Shapiro and Wilk's Test for Normality to Large Samples, *Journal of the Royal Statistical Society. Series C (Applied Statistics)*, Vol. 31, No. 2, 115–124, 1982.

[110] Ruben, H. Probability Content of Regions Under Spherical Normal Distributions, II: The Distribution of the Range in Normal Samples, *The Annals of Mathematical Statistics*, 31, No. 4, 1113–1121, 1960.

[111] Rupert, D., What Is Kurtosis?: An Influence Function Approach, *The American Statistician*, 41(1), 1–5, 1987.

[112] Ruxton, G. D and M. Neuhäuser. When Should We Use One-Tailed Hypothesis Testing? *Methods in Ecology and Evolution*, 1, No. 2, 114–117, 2010.

[113] R Core Team, *R Data Import/Export*, Version 3.5.0 (2018-04-23). https://cran.r-project.org/doc/manuals/r-release/R-data.pdf

[114] R Core Team, *R Language Definition*, Version 3.5.0 (2018-04-23) DRAFT. https://cran.r-project.org/doc/manuals/r-release/R-lang.pdf

[115] R Core Team, *R: A Language and Environment for Statistical Computing*. R Foundation for Statistical Computing, Vienna, Austria, 2018. http://www.R-project.org

[116] R Core Team, *foreign: Read Data Stored by Minitab, S, SAS, SPSS, Stata, Systat, Weka, dBase*, ... R package version 0.8–70, 2017. http://CRAN.R-project.org/package=foreign

[117] Sarkar, D., *lattice: Lattice Graphics*. R package version 0.20–35, 2017. http://CRAN.R-project.org/package=lattice

[118] Scheffé, H., *The Analysis of Variance*, John Wiley & Sons, New York, 1959.

[119] Schmid, C. F., *Statistical Graphics*, John Wiley & Sons, New York, 1983.

[120] Siegel, S., Nonparametric Statistics, *The American Statistician*, Vol. 11, No. 3, 13–19, 1957.

[121] Simkin, D. and R. Hastie, An Information-Processing Analysis of Graph Perception, *Journal of the American Statistical Association*, 82(398), 454–465, 1987.

[122] Shapiro, S. S. and M. B. Wilk, An Analysis of Variance Test for Normality (Complete Samples), *Biometrika*, 52, No. 4, 591–611, 1965.

[123] Sheskin, D. J., *Handbook of Parametric and Nonparametric Statistical Procedures*, 3rd ed., Chapman & Hall/CRC, Boca Raton, FL, 2003.

[124] Smyth, G. K., Australasian Data and Story Library (OzDASL), 2011. http://www.statsci.org/data

[125] Spence, I., No Humble Pie: The Origins and Usage of a Statistical Chart, *Journal of Educational and Behavioral Statistics*, 30(4), 353–368, 2005.

[126] Stewart-Oaten, A., Rules and Judgements in Statistics: Three Examples. *Ecology*, 76, No. 6, 2001–2009, 1995.

[127] Stoline, M. R. and H. K. Ury, Tables of the Studentized Maximum Modulus Distribution and an Application to Multiple Comparisons among Means, *Technometrics*, 21, No. 1, 87–93, 1979.

[128] Thode, H. C. Jr., *Testing for Normality*, Marcel Dekker, New York, 2002.

[129] Thompson, J. W., Coomb's Theory of Data, *Philosophy of Science*, Vol. 33, No. 4, 376–382, 1966.

[130] Tukey, J., Comparing Individual Means in the Analysis of Variance, *Biometrics*, 5, No. 2, 99–114, 1949.

[131] Tukey, J., *The Problem of Multiple Comparisons*, Unpublished Report, Princeton University, 1953.

[132] Tukey, J. W., *Exploratory Data Analysis*, Addison-Wesley, Reading, MA, 1977.

[133] Tukey, J. W., The Philosophy of Multiple Comparisons, *Statistical Science*, 6, No. 1, 100–116, 1991.

[134] Unwin, A., *Graphical Data Analysis with R*, The R Series, CRC Press, Boca Raton, FL, 2015.

[135] Venables, W. N. and B. D. Ripley, *Modern Applied Statistics with S*. 4th ed., Springer, New York, 2002.

[136] Venables, W. N., D. M. Smith and the R Core Team, *An Introduction to R. Notes on R: A Programming Environment for Data Analysis and Graphics*, Version 3.5.0 (2018-04-23). https://cran.r-project.org/doc/manuals/r-release/R-intro.pdf

[137] von Huhn, R., Further Studies in the Graphic Use of Circles and Bars; A Discussion of the Eell's Experiment, *Journal of the American Statistical Association*, 22(157), 31–39, 1927.

[138] Welch, B. L., On the Comparison of Several Mean Values: An Alternate Approach, *Biometrika*, 38, No. 3/4, 330–336, 1951.

[139] Wickham, H., *Advanced R*, Chapman & Hall/CRC, Boca Raton, FL, 2014.

[140] Wickham, H. *ggplot2: Elegant Graphics for Data Analysis*. Springer–Verlag, New York, 2009.

[141] Wolman, A. G., Measurement and Meaningfulness in Conservation Science, *Conservation Biology*, Vol. 20, No. 6, 1626–1634, 2006.

[142] Wuertz, D., Y. Chalabi, M. Maechler with contributions from J. W. Byers and others. *timeDate: Rmetrics—Chronological and Calander Objects*. R package version 3010.98, 2013. http://CRAN.R-project.org/package=timeDate

[143] Xie, Y. and M. Yu (2013). *rgabriel: Gabriel Multiple Comparison Test and Plot the Confidence Interval on Barplot*. R package version 0.7, 2013. https://CRAN.R-project.org/package=rgabriel

Index

Note: Boldface items represent R packages, objects/functions, and function argument names.

Printed and bound by PG in the USA